Molecular Photoelectron Spectroscopy

Molecular Photoelectron Spectroscopy

A Handbook of He 584 Å Spectra

D. W. Turner
Physical Chemistry Laboratory
University of Oxford

C. Baker
B.P. Chemicals, Epsom, Surrey

A. D. Baker
Chemistry Department, Swansea University

C. R. Brundle
Bell Telephone Laboratories, New Jersey

WILEY—INTERSCIENCE
a division of John Wiley & Sons Ltd
LONDON NEW YORK SYDNEY TORONTO

Library of Congress catalog card number 73-122351

ISBN 0 471 89320 X

Printed by offset in Great Britain by William Clowes and Sons, Limited, London and Beccles

Preface

During the past seven years Molecular Photoelectron Spectroscopy[1]* has been recognized as an especially unambiguous method for the study of the molecular electronic structures of substances in the vapour state. It is defined by the relationship:

$$E = h\nu - I_i - \Delta E_{vib} - \Delta E_{rot}$$

where $h\nu$ is the energy of a quantum of radiation from a monochromatic source. The rapid exploration of the field owes much to the use of the helium resonance line (548 Å equivalent to a photon energy of 21.2168 eV) as the light source, an innovation made at Imperial College, London.[2] The present volume presents some of the results of an extensive survey of the photo-electron spectra of many different classes of compound.[3-7]

Many of these spectra are of substances studied in the earlier phases of our work and have been re-recorded on apparatus of improved performance but the original critical discussion and the interpretation given by earlier colleagues M. I. Al-Joboury, T. N. Radwan and D. P. May in their Ph.D. Theses (London University 1964, 1965 and 1966 respectively) is in many instances still relevant and has frequently been used as a basis for a more detailed account.

We seek in this volume to draw the attention of research workers in chemistry to the main features of the spectra of a range of substances of vary-ing degrees of molecular complexity. Not all of the substances whose spectra we have recorded have been included in the discussion. Neither have we attempted to account for every detail in the spectra of many of the compounds which we have selected for discussion. The motive for producing this volume

---

* References quoted in the Preface can be found on page 14.

is the feeling that our good fortune in acquiring a large mass of previously inaccessible data leaves us with an obligation to collect this into one place for ease of access even before a complete account of its interpretation is possible.

Acknowledgements

We have been guided in the interpretation of our spectra by discussion with so many spectroscopists and quantum chemists that we cannot attempt to acknowledge all here individually. To both Professor W. C. Price, F.R.S., King's College, London, who was among the first to recognize the significance of this technique especially in relation to his own pioneering work on Rydberg series in polyatomic molecules and Professor E. Lindholm, Royal Technical Institute, Stockholm, who was quick to give us the benefit of his long experience of the elucidation of electronic structure of complex molecules by mass spectroscopy, however, we give our especial thanks.

Chapter 8 owes much to the development by Dr. A. Orchard of group theoretical procedures for the interpretation of the photoelectron spectra of species where spin orbit coupling is large. Part of this chapter has been rewritten under his guidance for which we are extremely grateful.

The construction of the high resolution photoelectron spectrometer mentioned in the text was made possible by a grant from the Paul Instrument Fund of the Royal Society and was constructed by the Imperial College Workshop (Mr. F. C. Cobley) and by Mr. T. F. Adey who are warmly thanked. The Science Research Council provided other financial support including research studentships (to A.D.B., G.R.B. and C.B. and to Dr. D. P. May).

Many of the compounds whose spectra we recorded here were provided by colleagues in other institutions often with the object of elucidating a point of electronic structure (usually the first ionization potential) distinct from our own interest in these compounds. In a number of cases joint publications have resulted and are cited in the text. The following are thanked in this connection for generous gifts of material and fruitful discussions: B. J. Atkinson, J. Callomon, J. Collin, J. Delwiche, G. B. Ellison, J. A. Elvidge, S. Evans, J. Green, M. L. H. Green, G. King, P. Joachim, R. G. Luckcock, J. C. Maine, M. Robin, A. Smith, K. B. Wiberg, G. R. Wilkinson.

The work reported here had its beginnings in the organic chemistry laboratory at Imperial College, London, and we are especially indebted to Professor D. H. R. Barton, F.R.S., for his encouragement and support.

D. W. TURNER

Balliol College, Oxford
September 1969

Contents

Contents

Chapter 12   Heterocyclic Aromatic Compounds

Chapter 13   Hydrogen Cyanide and Related Compounds

Chapter 14   Miscellaneous Inorganic Compounds

CHAPTER 1

Introduction

1. FORMAT OF THE SPECTRA

Each spectrum is a continuous record of electron count rate, integrated over either 0·33, 1 or 3 seconds, plotted as a function of electron kinetic energy (see Chapter 2). In the absence of a recognized convention for the plotting of photoelectron spectral curves we have adopted scales which relate most closely to the primary data. The ordinate scale of electron flux is in count sec^{-1} throughout and increases upwards and the abscissa scale of electron energy (E) increases from left to right. This latter choice requires some defence for the secondary, deduced data, ionization energy ($I = hv - E$) then increases from right to left contrary to *inter alia* the convention in mass spectroscopy literature. The arrangement we have chosen however is consistent with the English spectroscopy usage where wavelength increases from the left and so makes for easy comparison between photoelectron spectra and conventional optical spectra.

In all the examples given a full spectrum is presented which extends from zero electron energy (I.P. 21·2168 eV) to 10 or more eV so as to include all the observed bands though some may omit a section near zero kinetic energy if no band is found here in order to fit the remainder of the spectrum to a convenient range. Many of the spectra were obtained with a recording system[8] which covered a maximum range of 10 eV at one setting and in some cases this has necessitated the joining of two spectra together. When this has occurred a break is indicated in the spectrum but no omission of detail in the join need be feared as the region of overlay was always recorded separately to avoid this possibility.

Sections to Expanded Scales

It has frequently been found that the subtleties of vibrational fine structure cannot be seen on the full scale presentation especially where only partially resolved details and changes in line shape are of significance. The resolution of detail in simple molecules may be of the order of the best instrumental resolution—about 15 mV—and in such cases the appropriate regions were rerecorded with (usually) a 5- or 10-fold expansion of the abscissa scale and at a lower speed and increased integration time. This also allowed more accurate calibration of the energies of the more prominent spectral features.

In a similar manner it has often been necessary to record again to a different ordinate-scale sections where the electron flux has been low. This might arise from unusually low ionization cross-section but more commonly is due to great differences in Franck–Condon factors (see below p. 7) between one band and another making it difficult to discern all the details on different portions recorded to the same scale. In those spectra recorded using a scanning system based on variation of the analyser field (which is generally the case—see below) the electron flux obtained near the beginning of the spectrum (low electron energy) is also low for the effective spectral slit-width increases with energy under these conditions.

The factors by which the ordinate scale has been changed are included where they are exact factors determined by ratemeter range switching for example but in many cases the increase in sensitivity was an arbitrary one and must be estimated directly from the record.

A number of spectra have been recorded with a fixed analyser-plate potential difference, the spectrum being scanned by variation of the potential of an accelerator electrode within the target chamber. The analyser was set to bring 6 eV electrons to a focus so that the accelerator-electrode bias varied from -6 V to $+4$ V to cover the basic 0–10 eV range. Such spectra bear the code letter [R]. In this arrangement the analyser had an effectively constant spectra slit-width, and the usually large flux of nearly-zero energy electrons observed is a possible drawback. Further, in some cases, additional broad bands were sometimes observed in the 0–3 eV region, suggesting that electron–molecule collision processes might be of significance when an appreciable accelerating field is present. When some electrons of already high kinetic energy are present—as, for example, with benzene in the target chamber (K.E.$_{max}$ $\approx$ 12 eV)—the increase to 15–18 eV may be high enough to give rise to secondary excitation or even ionization processes.

We have included for completeness a number of spectra obtained using the magnetically focussed analyser of Turner and May[5,9]. This instrument possessed the advantage of a rather high collecting power stemming from its small electron orbit radius, large slits (0.1 mm × 1 cm) and intimate juxtaposition of ionizing region and analyser entrance slit. While the spectra were in consequence rather poorly resolved they could be readily obtained for compounds of lower volatility than could be easily studied by apparatus of better resolving power but smaller collecting power. In most cases the amount of information lost by not employing the highest resolution was small since all those examples were of molecules of such complexity and low symmetry that only featureless bands free of resolvable vibrational fine structure were expected (or indeed usually found when they have been repeated using better instruments). We feel therefore that their inclusion is justified on the grounds that much new information is contained in these spectra and it is likely to suffer only small modification in the light of subsequent and possibly only slightly better instrumental records. These spectra bear the code letter [M].

One inherent disadvantage of the magnetic spectrometer which must be borne in mind is that it is basically a momentum scanning device. The abscissa scale is not therefore linear in energy though in the present case it is intermediate between this and linearity in momentum as a result of the characteristics of the scanning arrangement. In this the abscissa value was read as a linear function in potential by a 'read-out' potentiometer which controlled a current regulator feeding the Helmholtz coils of the analyser. This current regulator had a nonlinear characteristic.

2. PHOTOIONIZATION AND THE FORMATION OF A PHOTOELECTRON SPECTRUM

Photons in the vacuum ultraviolet region of the spectrum whose energy is about 10 eV interact with atomic or molecular gases mainly to cause excitation that is to promote electrons to other bound states even though there may be sufficient energy to cause ionization. The careful study of absorption spectra has however afforded many examples of regions of continuous absorption such as might be ascribed to immediate ejection of a free electron.[10] Such a process has an oscillator strength f approaching unity but energy conservation with a wide range of possible electron energies permits the photon absorption process to occur over a wide wavelength range. This usually results in the cross-section for the ionization at any particular wavelength being small compared with cross-sections for excitation where the whole oscillator strength is concentrated in much shorter wavelength range. Since continuous absorption may also arise from dissociation into molecular fragments or atoms;

$$M \xrightarrow{h\nu} A\cdot + B\cdot$$

the positive identification of the limits of a true ionization continuum is often uncertain.

Watanabe[11] partially overcame the confusion between the many different modes of absorption by restricting measurements to those in which only charged species were produced i.e.

$$M \xrightarrow{h\nu} A^+ + B^- \tag{1}$$

$$\text{or} \quad M \xrightarrow{h\nu} M^+ + e^- \text{ (initially)} \tag{2}$$

Though no distinction between these two processes is afforded the latter is usually the most common. The technique (Figure 1.1) consists of varying the wavelength of the light from a monochromator and observing the onset of the production of charged species in irradiated vapour by measuring the total current which can be passed between a pair of electrodes spanning the

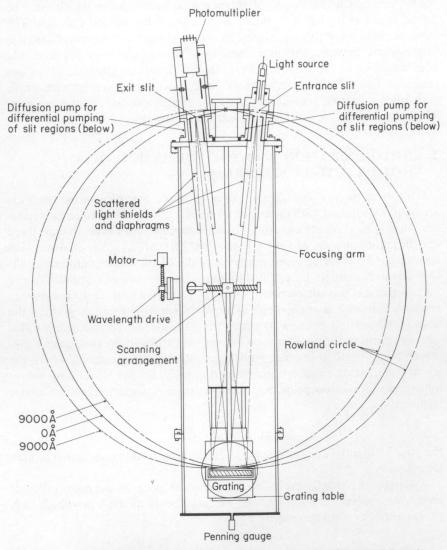

Figure 1.1   Vacuum monochromator and ionization chamber for measuring photo-ionization cross-section. (M. A. Hourieh and D. W. Turner unpublished work, M. A. Hourieh Ph.D. Thesis, London University. 1965)

illuminated region. It is instructive to compare the photoionization yield curve and absorption spectrum obtained for N_2 by Cook and Ogawa[12] which are shown in Figure 1.2 over the same energy scale. The sudden increase of current from zero at a critical wavelength marks usually, but not exclusively, the minimum ionization potential for electron emission.

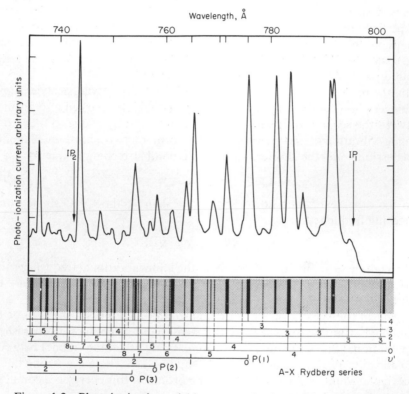

Figure 1.2 Photoionization yield-curve and absorption spectrum for Nitrogen gas. Obtained by Cook and Ogawa, reproduced from Reference 12 by permission

In photoelectron spectroscopy we concern ourselves only with the processes which liberate electrons

$$M \overset{h\nu}{\longrightarrow} M^+ + e^- \qquad \text{(Direct ionization)} \qquad (3)$$

and possibly

$$M \overset{h\nu}{\longrightarrow} M^* \longrightarrow M^+ + e^- \qquad \text{(Autoionization)} \qquad (4)$$

where M^+ may be in a stable, metastable or unstable (repulsive) state.

3. ENERGY LEVEL DIAGRAMS

Inspection of the photoelectron spectra for compounds that are members of a series related by common electronic structural features (for example iso-electronic series) frequently reveals marked resemblances particularly in the shapes of selected bands. That is to say the Franck–Condon envelopes of bands arising in the ionization of electrons from orbitals whose symmetry properties, and LCAO coefficients provide by their apparent invariance a means of identification. This takes on greater precision when the vibrational fine structure exhibits demonstrably common features as between the same band observed in the spectra of different substances.

In one particular band of the spectra of diacetylene, cyanoacetylene and cyanogen for example (Figures 6.23, 6.26 and 6.27) the marked resemblance between the second, third and third bands respectively is due to their containing vibrational fine structure in the form of two short, overlapping progressions one, the shorter, in the triple bond stretching frequency v_1:

$$\leftarrow X{\equiv}C \rightarrow\!\!-\!\!\leftarrow C{\equiv}X\rightarrow \qquad (X = N \text{ or } CH)$$

and the other extending up to about $v' = 4$ in v_2 the C—C single bond stretching frequency:

$$\leftarrow X{\equiv}C \longleftrightarrow C{\equiv}X\rightarrow$$

Though there are minor differences in the ionic vibrational frequencies the strong general resemblance suggests a common assignment, in this case to ionization from the deeper of two π levels.

Again in the spectra of many organic compounds possessing a 5- or 6-membered aromatic nucleus there is a small band sometimes distinguished by vibrational fine structure (? the symmetrical 'breathing' mode) between 16·5 and 17·5 eV (cf. Figures 10.2, 12.12 and 12.19). There is a strong inference that a symmetric, ring-atom binding orbital at this energy is a common feature of such compounds.

With such internal evidence of common bond assignments it is possible to compare the data from related series of compounds in a very straight-forward manner by drawing energy level diagrams based (preferably) on the vertical ionization energies deduced from the positions of the peaks of the bands.

Such diagrams are shown in Figures 3.19, 4.20, 4.42, 5.22, 6.34 and 7.7.

These may be likened to orbital energy level diagrams. They could only have such a strict equivalence however on a basis of rigid adherence to the precepts of Koopmans' Theorem.[13] This postulates equality of the negative of an ionization energy to a one electron orbital energy. Since this takes no account of the contribution to the ionization energy by reorientation of the remaining electrons it is clearly in error but to an extent which though uncertain in most cases seems to be of the order of 1–2 eV and possibly to differ from one orbital to another.

The energy level diagrams that we find can be seen to be strictly speaking ionic energy level diagrams showing the relative energies of different ionic configurations compared to the ionic ground state. These would normally be plotted upside down, excited configurations then appearing above the ground state configurations. We have nevertheless adhered to the form of a conventional molecular orbital energy level diagram (and ignore by implication deviation from Koopmans' Theorem) on the grounds that in this way the salient features associated with for example electro negativity changes on the introduction of heteroatoms have immediate chemical significance.

4. THE FRANCK–CONDON PRINCIPLE IN IONIZATION

According to the Franck–Condon principle the transfer of excitation in an electronic transition occurs in a time which is short compared with that required for the execution of vibrational motion. It governs therefore in the present instance the relative probabilities of ionizing transitions from (usually) the molecular vibrational ground state ($v'' = 0$) to the various ionic vibrational states ($v' = 0, 1, \ldots$).

In the approximation of a transition moment, $G_{e'e''}$, which varies slowly with internuclear coordinates, r, the probability $P_{v'v''}$ is given by the Born–Oppenheimer equation:

$$P_{v'v''} = [G_{e'e''}]^2 \left[\int \psi_{v'} \psi_{v''} \, dr \right]^2$$

The term in the second bracket is referred to as the Franck–Condon factor and its square is directly proportional to $P_{v'v''}$ if $G_{e'e''}$ is constant.

The electron flux for each peak in a photoelectron spectrum showing resolved vibrational structure measures either directly or indirectly (see below) $P_{v'v''}$ at a particular energy (here 21·2 eV). Since for different vibrational levels v' the excess energy $E \, (= h\nu - I - E_{vib})$ is different $G_{e'e''}$ will not be a constant. It may however be expected to be only slowly varying where direct ionization is occurring. The array of peaks in the photoelectron spectrum is thus approximate in amplitude to the square of the Franck–Condon factors and their envelope to the so-called 'Franck–Condon envelope'.

A retarding-field energy analyser is capable of giving such information directly since it is equally sensitive to slow or fast electrons. Focusing deflection analysers with fixed slits in which the spectrum is scanned by varying the focusing field have a 'spectral slit-width' which increases with electron energy (electrostatic field) or momentum (magnetic field). It is necessary here therefore to correct the peak heights for this instrumental effect to obtain the Franck–Condon factors.

To see exactly how the Franck–Condon principle and the Born–Oppen-

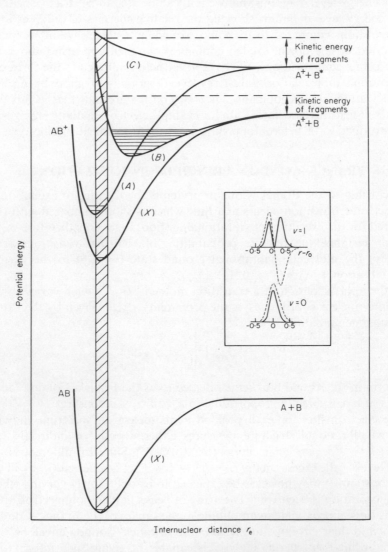

Figure 1.3   Potential energy curves for a diatomic molecule **AB** in its ground state (X) and for its ion $\mathbf{AB}^+$ in its ground state $(\tilde{X})$ and three excited states $(\tilde{A})$ $(\tilde{B})$ $(\tilde{C})$ (schematic). Inset, vibrational wave-functions (dotted) and probability distributions (full lines) of an anharmonic oscillator for $v = 0,1$ (from Reference 1.14, p. 94)

heimer approximation affect electronic transitions refer to Figure 1.3. Transitions are all taken as originating from the zeroth vibrational level of the initial electronic state. In this vibrational level $\psi_{v''}$ has its maximum at the equilibrium position, so the maximum probability for transition occurs in the centre of the Franck–Condon region (marked in Figure 1.3 by shading). Significant transitions cannot occur outside the Franck–Condon region ($\psi_{v''}$ approaches zero), and how far away from the centre they can be detected depends on the sensitivity of measurement.

Excitation to the electronic state $(\tilde{X})$, where the equilibrium internuclear distance, r'_e, is the same or very nearly the same as that in the ground state, (X), requires by the Franck–Condon principle that the most probable transition is to the $v' = 0$ level of state $(\tilde{X})$. Transition to other vibrational levels will have a very low probability which decreases as v' increases. Transitions to the $v' = 0$ level are termed ADIABATIC TRANSITIONS.

Excitation $(\tilde{A}) \leftarrow (X)$ corresponds to the case where r'_e is slightly less than r''_e. From the diagram it can be seen that the maximum overlap of wave functions occurs at $v' = 1$ and thus the maximum transition probability occurs at this level. Transitions will occur to other levels, the probability of which depends on $[\int \psi_{v'} \psi_{v''} \, d\tau]^2$. The transition of maximum probability is termed a VERTICAL TRANSITION. In the case of excitation $(\tilde{X}) \leftarrow (X)$ the adiabatic and vertical transitions are identical. In the case of $(\tilde{A}) \leftarrow (X)$ they are not. The adiabatic transition is strictly to the $v' = 0$ level, but transitions to this level may be of a very low probability if the difference between r'_e and r_e is large (e.g. excitation $(\tilde{B}) \leftarrow (X)$). Hence the experimental adiabatic transition is then the one of lowest energy which is detectable.

Both transitions $(\tilde{X}) \leftarrow (X)$, and $(\tilde{A}) \leftarrow (X)$ represent a case—(1) excitation process (Section 3, Chapter 1), and could equally well represent case (7), if the transitions were to molecular, not ionic levels.

Transition $(\tilde{B}) \leftarrow (X)$ shows the situation where there is a very large change in r_e. Here transitions occur to the level of the continuum and the ion fragments into $A^+ + B$. The fragments will possess kinetic energy but there will be a spread in this energy reflecting the shape of $v' = 0$.

Transition $(\tilde{C}) \leftarrow (X)$ represents a transition to an unstable state and fragments possessing kinetic energy will against result.

Thus all these types of *direct* ionizing transition are possible in a molecule A—B, the transitions being governed by the Franck–Condon principle.

Correlations between Franck–Condon Factors, Changes in Vibrational Frequencies, and the Bonding Type of the Electron excited in Transitions

It is interesting to see how the vibrational frequencies changed on transition to each of the ionic states $(\tilde{X})$, $(\tilde{A})$, $(\tilde{B})$ and $(\tilde{C})$.

The energies of the vibrational levels are given by

$$E_v = (v' + \tfrac{1}{2})h\omega - (v' + \tfrac{1}{2})^2 hx\omega + \text{smaller terms}$$

where v' = vibrational quantum number in the ion

h = Planck's Constant

x = anharmonicity constant

$\omega = \sqrt{k/u'}/2\pi$, the vibrational frequency

k = force constant of the vibration

u' = reduced mass of the system.

Now k is a measure of the strength of the bond. If a nonbonding electron is removed on ionization, the bond strength will alter little, and so k, and hence ω and the energy separation between vibrational levels, will remain almost unchanged. The bond length r_e will also be unaffected. On removing a bonding electron, k and ω will decrease, and r_e will increase. Conversely, on removing an antibonding electron, k and ω increase and r_e decreases. Similar effects occur when the electron is just excited to a higher molecular level and not completely removed.

Thus in general, from Figure 1.3, it can be said that:

(1) $(\tilde{X}) \leftarrow (X)$ ionizing transitions correspond to the removal of a nonbonding electron, with the vibrational frequency, ω', in state (X) being essentially the same as ω in state (X).

(2) $(\tilde{A}) \leftarrow (X)$ corresponds to the removal of an antibonding electron, and the vibrational frequency in state $(\tilde{A})$ is increased somewhat.

(3) $(\tilde{B}) \leftarrow (X)$ results from the removal of a very strongly-bonding electron, the vibrational frequency being greatly reduced.

Since the values of the Franck–Condon factors are dependent on r_e they will also be characteristic of the bonding type of electron removed. Removal of nonbonding electrons leads to transition essentially to the $v' = 0$ level, while bonding or antibonding electrons will significantly populate higher v' levels, so that the maximum Franck–Condon factor may not be that for $v' = 0$. The argument above can be extended to polyatomic systems but in practice it becomes complex because even for linear triatomic molecules three vibrational modes are involved and a three-dimensional potential surface is required to describe correctly the vibrational motion of the molecule.

It is of some interest nevertheless to see whether or not some practical use may be made of the vibrational fine structure which is observable even for quite large polyatomic molecules.

In optical spectroscopy there are a number of well-known empirical relationships linking the internuclear distance r_e and the vibrational frequency (*a*) in different electronic states of the same molecule and (*b*) in a related series of molecules. The reader is referred to Herzberg[14] for a more complete discussion of the various 'rules'. We may extend case (*a*)—different electronic state of the same molecule—to include the ionic states[15] using Mecke's relationship:

$$\omega_e \propto \frac{1}{r_e^{\,2}}$$

For small changes in r_e we may assume the potential function to be approximately parabolic so that the difference between the adiabatic and vertical

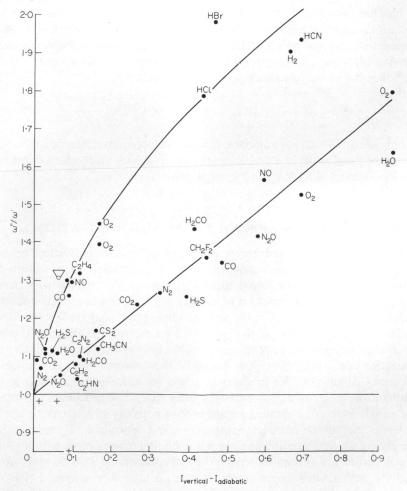

Figure 1.4 Plot of the ratio of vibrational frequencies in molecule and ion (ω''/ω') versus the difference between the Adiabatic and Vertical Ionization Potentials

ionization energies $I_V - I_A$ which is in general a function of the fractional change in internuclear distance may be expressed as:

$$I_V - I_A = f\left(\frac{r'}{r''}\right)^2$$

Hence from Mecke's relation we expect:

$$I_V - I_A = f\left(\frac{\omega''}{\omega'}\right)$$

This has been tested using examples about whose vibrational assignments there is no ambiguity and the results are summarized in Figure 1.4. It would appear that though there is in fact some agreement with the expected linear relationship some grouping near two curves is apparent. For the main branch the following linear relationship is approximately true:

$$I_V - I_A = 1\cdot2\left(\frac{\omega''}{\omega'} - 1\right)$$

From measured vertical adiabatic ionization potential difference therefore it may be possible in favourable cases to infer which molecular vibrational mode an observed ionic frequency represents.

5. PERTURBATIONS ARISING FROM AUTOIONIZATION

The arguments outlined above represent an oversimplification on two counts. Firstly it has been assumed that the ionization cross-section for production of each vibrational ionic state varies only slowly with excess energy E and that this function is independent of the vibrational quantum number v'. Implicit in this is the second assumption that the only mode of ionization of significance is that in which an electron is liberated directly upon absorption of the photon energy. That is to say the electron enters the ionization continuum directly. It is well known however that autoionization processes which give rise to sharp maxima (or sometimes minima) in the total ionization cross-section curve for both atoms and molecules provide at the wavelength of the maxima an alternative pathway which may be several orders of magnitude more probable than direct ionization. Therefore when the excited state which autoionizes has a lifetime comparable with a vibrational period (as is implied where sharp resonances occur) to give vibrationally well defined upper states, it can no longer be assumed that the electron is lost from a nascent ion whose internuclear separation is the same as that in the molecule at the moment when the photon was absorbed. Here, therefore, we expect that the probability for eventual production of the various ionic vibrational levels will reflect the Franck–Condon factors connecting them to the autoionizing excited state rather than the ground state.

Several examples of this effect, which was first demonstrated by W. C. Price,[16] have now been found. Oxygen shows a marked difference in the vibrational envelope of the photoelectron band for the $^2\Pi_g$ ionic state when neon resonance-line photons ($hv = 16\cdot8$ eV) are used. A marked enhancement of the vibrational levels above $v' = 5$ is found. Similar effects have been reported for nitrogen and carbon monoxide by Collin.[17] An example from the spectrum of carbon disulphide (Figure 1.5) is shown here. The $v' = 1$ level in the ion is enhanced by a factor of about 100 by the first argon resonance line ($hv = 11\cdot84$ eV) but by a much small factor by the second ($hv = 11\cdot62$ eV).

Though the effect may be observed only over very small energy ranges and so far only at the lower photon energies we should not assume that in any particular case it cannot occur for $hv = 21\cdot21$ eV. It may be therefore that certain features in the photoelectron spectra recorded here, distortions in the

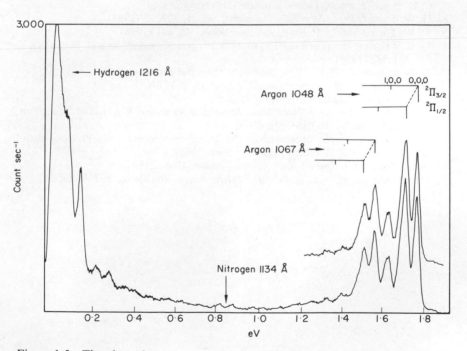

Figure 1.5 The photoelectron spectrum of carbon disulphide obtained when argon is added to the helium discharge. Abscissa, electron kinetic energy. Groups of photoelectrons ascribable to ionization by the Argon 1048 Å and 1067 Å, the Nitrogen 1134 Å and the Hydrogen 1216 Å line are indicated. The 1,0,0 level of the $^2\Pi_{3/2}$ state of the ion is unusually populated by Ar 1048 Å ionization

normal smooth Franck–Condon envelopes for example may be ascribable to this effect. Some marked 'double maxima' found in cyclopropane, allene, methane and benzene for example though reasonably ascribable to Jahn–Teller forces may in reality have their origin in the effects described above. Only studies with alternative radiation sources can eliminate this possibility completely. The helium $I\beta$ line 1·9 eV to higher energy than the 21·21 eV α line may be useful for this purpose.

REFERENCES

1. M. I. Al-Joboury and D. W. Turner, *J. Chem. Soc.*, **1963**, 5141; F. I. Vilesov, B. C. Kurbatov and A. N. Terenin, *Dokl. Akad. Nauk SSSR*, **138**, 1329 (1961).
2. D. W. Turner and M. I. Al-Joboury, *J. Chem. Phys.* **37**, 3007 (1962).
3. M. I. Al-Joboury, *Ph.D. Thesis*, London University (1964).
4. T. N. Radwan, *Ph.D. Thesis*, London University (1966).
5. D. P. May, *Ph.D. Thesis*, London University (1966).
6. A. D. Baker, *Ph.D. Thesis*, London University (1968).
7. C. K. Brundle, *Ph.D. Thesis*, London University (1968).
8. D. W. Turner, *Proc. Roy. Soc. (London)*, *Ser. A*, **307**, 15 (1968).
9. D. W. Turner and D. P. May, *J. Chem. Phys.*, **45**, 471 (1966).
10. G. L. Weissler, *Handbuch der Physik* (Ed. Flügge, Springer Verlag), Berlin XXI, pp. 304–382 (1956).
11. K. Watanabe, *J. Chem. Phys.* **26**, 542 (1959), and earlier papers.
12. G. K. Cook and M. Ogawa, *Can. J. Phys.*, **43**, 256 (1965).
13. T. Koopmans, *Physica*, **1**, 104 (1934).
14. G. Herzberg, *Molecular Spectra and Molecular Structure*, Vol. 1, 2nd Edn., Van Nostrand, New York, 1954, pp. 456 ff.
15. D. W. Turner, Paper presented at the Discussion meeting on Photoelectron Spectroscopy, Royal Society, London, Feb. 1969.
16. W. C. Price, *Mol. Spectr. Rept. Conf.*, *London*, **1969**, 221.
17. J. E. Collin and P. Natalis, *Intern. J. Mass Spectr. Ion Phys.*, **2**, 231 (1969).

CHAPTER 2

Experimental Methods

1. GENERAL ARRANGEMENT

Two earlier experimental arrangements[1,2] are illustrated in Figures 2.1 and 2.2. The distinctive feature of the apparatus described by Al-Joboury and Turner[1] (Figure 2.1) is a helium resonance light source A, separated from the ionization chamber C by two aligned sections of precision bore (0·5 mm) Pyrex capillary tube, B. The capillary tube sections prevent the target vapour from penetrating into the light source without the necessity for a window and also provide a narrow well-collimated photon beam which could be aligned

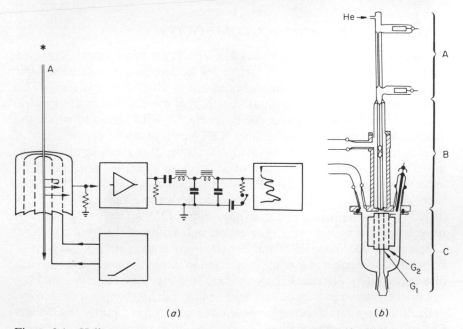

(a)   *(b)*

Figure 2.1   Helium resonance photoelectron spectrometer of Al-Joboury and Turner. (a) General arrangement; (b) cross-section of light source A, collimator section B and target chamber C

with the axis of the electron-retarding grid structure G_1, G_2. It will be shown below that this arrangement contributes markedly to the energy resolution attainable.

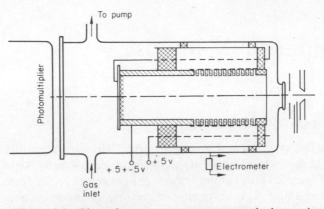

Figure 2.2   Photoelectron spectrometer attached to exit-slit of a vacuum monochromator (after Vilesov and colleagues, Reference 1.1)

2. LIGHT-SOURCE MONOCHROMATICITY

Under the appropriate conditions, the helium emission spectrum emitted by a d.c. discharge consists of the series leading to the ionization limit (24·47 eV) of which by far the strongest line is He 584 Å (21·22 eV). This resonance line then accounts for at least 98% of the emission in this spectral region. At longer wavelengths, we have the $^3P \rightarrow {}^3S$ series, of which the shortest wavelength emission is 3000 Å ($\sim$4 eV), followed by a number of lines in the visible, notably $\lambda = 5875$ Å (yellow). It is clear, therefore, that in the vast majority of substances, which have an ionization potential (I.P.) of 5 eV or greater, essentially the only ionization caused is due to the resonance line He 584 Å. Traces of hydrogen, which are removed only with difficulty, cause emission of the Lyman α line 1215 Å (10·20 eV) and both oxygen and nitrogen, readily released on outgassing of the lamp structure, have a many-line spectrum of wavelengths shorter than 1000 Å. Such emissions are also capable of exciting ionization in most substances and can confuse the resultant electron energy spectra. It is essential, therefore, that the gas in the discharge tube is extremely pure. Fortunately, this is not too difficult using a liquid-nitrogen cooled charcoal trap to purify the helium. The state of the discharge is readily assessed by examining the visible spectrum for absence of the hydrogen Balmer series and oxygen and nitrogen visible emission lines. When operating correctly, the light source is a yellowish

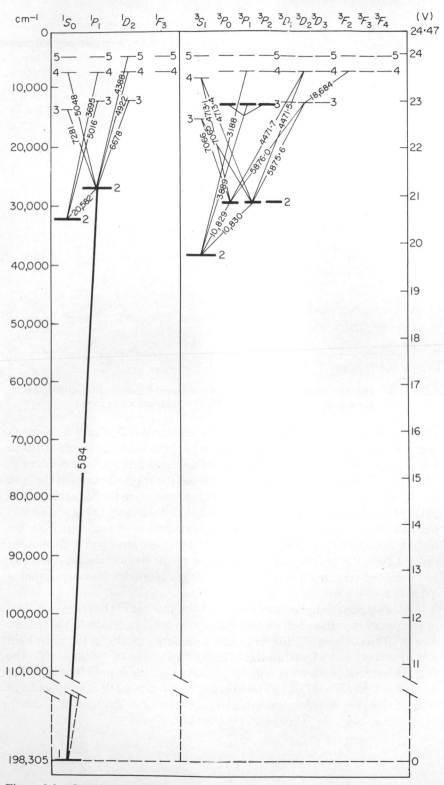

Figure 2.3   Grotrian diagram for Helium showing the wavelengths of the strongest emissions

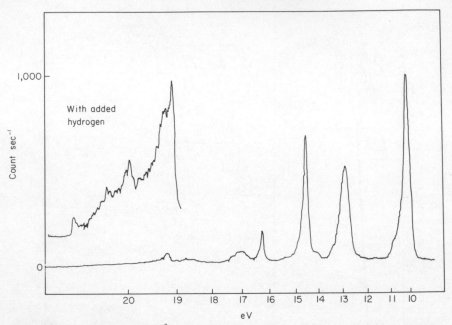

Figure 2.4   The Helium 584 Å photoelectron spectrum of carbon disulphide showing
the effect of adding hydrogen to the light-source gas [M]

peach colour, free from blue around the electrodes. Features in electron
energy spectra which arise from ionization by those lines due to impurities in
the discharge tube are easily recognized, however, as they increase in intensity
relative to the remainder of the spectrum when the spectral purity of the light
source deteriorates. For example, a weak but sharp line in the carbon disulphide photoelectron spectrum, earlier assigned tentatively to the sixth I.P.,[3]
has now been proved[4] to arise from photoionization of electrons from the
highest occupied level (I.P. = 10·11 eV*) by hydrogen Lyman β photons
($h\nu = 12·09$ eV), since on adding hydrogen to the helium stream, the weak
line increases very much in intensity while the remainder of the spectrum is
quenched.

It has also been found possible to make positive use of the resonance lines
of elements other than helium by adding a small proportion to the helium
flow.[5, 6] Thus, adding hydrogen in small amounts results in the main light
output becoming the Lyman α line of atomic hydrogen ($h\nu = 10·20$ eV). The
effect in benzene is shown in Figure 2.6. Argon yields mainly the resonance
line at 1067 Å ($h\nu = 11·62$ eV) (see Figure 2.7), but in both cases quite large
intensities reside in other lines than the desired one, the Lyman β line in

* Mean of doublet, spectroscopic values 10·08, 10·13 eV.

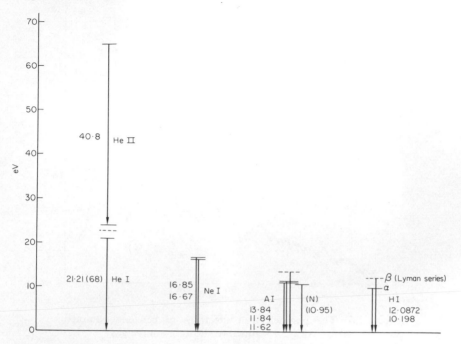

Figure 2.5   The principal resonance lines useful in photoelectron spectroscopy. The line marked *N* is of an impurity, possibly atomic Nitrogen of Xenon which may be present in an argon discharge

hydrogen and the other resonance line in argon and in impurities. Helium remains the most generally satisfactory radiation source.

By employing much higher current density (increased about ten-fold) and rather lower helium pressure Price[7] has shown that it is possible to detect photoelectron spectra originating from the 303 Å line of ionized helium (He II). The 584 Å line is still dominant however and the ionization energy range between about 30 eV and 40·8 eV is usually obscured by the overlying bands arising from ionization of the outer shells by the 584 Å radiation. Even so Price has been able to detect in a number of important cases energy levels that lie just below 21·21 eV notably those in the diatomic molecules nitrogen, nitric oxide, carbon monoxide and oxygen and in methane and ethane.

Controlled excitation of such higher resonance lines by high voltage electron beams in low pressure gases seems to offer hope of sources in which the contribution from the longer wavelengths would be much reduced. One possible arrangement is shown in Figure 2.8 where an electron beam originating in a low pressure region is focused through an aperture where it enters a

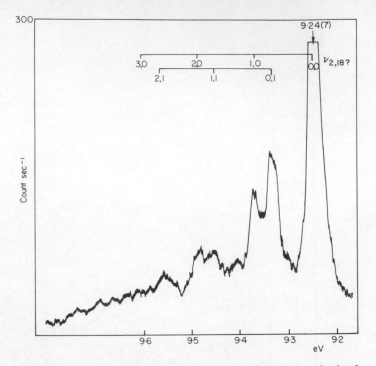

Figure 2.6   The photoelectron spectrum of benzene obtained
when hydrogen is added to the light-source gas

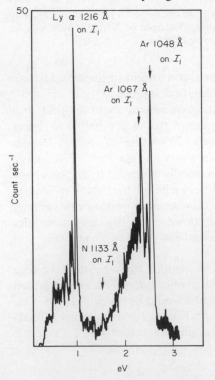

Figure 2.7   The photoelectron spectrum of benzene obtained when Argon
is added to the light-source gas (note
the presence of hydrogen as one
impurity)

region of higher pressure so that excitation of this gas occurs at a well defined electron energy and over a very limited volume.

Filament supply

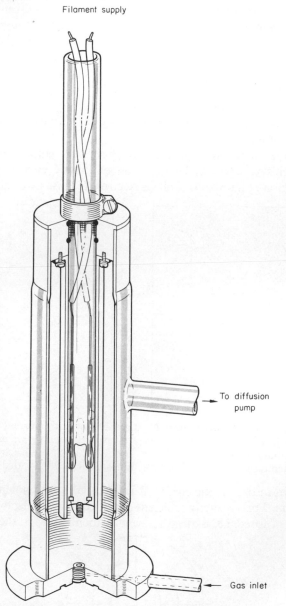

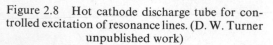

Figure 2.8   Hot cathode discharge tube for controlled excitation of resonance lines. (D. W. Turner unpublished work)

3. PHOTOELECTRON ENERGY ANALYSIS

The methods for determining the kinetic energy spectrum of a 'polychromatic' beam of electrons are many and varied, ranging from the simplest, which employ merely electrostatic retarding fields, through deflexion analysis by means of radial electrostatic or magnetic fields and combinations of these, to highly chromatic lens systems. The subject has been recently reviewed by Klemperer[8, 9] who describes many examples in detail. Some of these designs are, however, inherently unsuited to the present task owing to their low collection efficiency for an electron flux emerging radially from a line source.

Both of the early photoelectron spectrometers employed electrostatic retarding fields to analyse the spectrum of kinetic energies obtained on photoionization (Figure 2.9). For an electron to pass through two grids, it

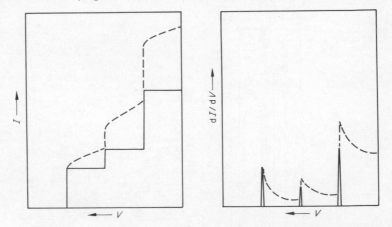

Figure 2.9   Idealized form of curve of collector current I versus retarding potential V, and its first derivative the photoelectron energy-distribution function. The effect of deviations of the electron paths from the electric vector direction is indicated by dotted lines

must have an initial kinetic energy at least equal to the potential difference between them. This is exactly sufficient provided that the electron is travelling along a line of force, i.e. is always normally incident upon the equipotential surfaces.

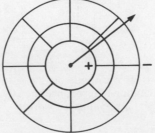

To make use of as much of the photoelectron flux as possible, cylindrical grids were used for which this condition of normal incidence applies for electrons emitted radially from the axis. The electron energy spectrum is obtained as the first derivative with respect to retarding field of the electron current emerging from this second grid to a suitably biased collector electrode (Figure 2.1). The differentiation can be done manually by point-by-point plotting of the current changes following a succession of small potential increments, the technique used by Vilesov and Kurbatov[2] and subsequently by Schoen.[10] Alternatively, a continuous spectrum can be recorded by applying a potential sweep which is linear with time and simultaneously recording the first derivative of the output current, as was described by Turner and Al-Joboury.[1] An elegant variation on this method has been described by Comes who divided both retarding grid structures into two parts with a small difference dV in the retarding field between them. Using two collector electrodes to receive the separate currents I_1 and I_2, the difference in these currents I_1 and I_2 was recorded as proportional to dI/dV. Some such continuous recording technique is to be preferred, especially when fine structure is present in the spectrum but it introduces some problems in obtaining an adequate signal-to-noise ratio. Probably it cannot be applied with the very small signals obtained when a monochromator is used as a light source.

4. ENERGY RESOLUTION AND THE FORM OF THE SPECTRUM FROM GRID SPECTROMETERS

The resolving power is readily tested by examining the electron energy spectra of the rare gases, which give simple doublets because spin–orbit splitting distinguishes the two states of the ion $^2P_{3/2}$ and $^2P_{1/2}$ (Figure 2.10).

The ultimate sharpness of the lines in the photoelectron energy spectrum will be determined by the sum of the Doppler broadening for the helium atoms and target gas molecules (about 0·2 an 0·05 mV, respectively, at room temperature) and by the contribution to the photoelectron velocity from thermal motion of the target molecules (about 2 mV in cases of practical interest).[11,12]

By careful choice of mesh spacing in relation to grid separation, by gold coating the electrodes, and by excluding all insulating materials (which can acquire surface charges and consequently cause field distortion) from between the grids, it has been possible to achieve a resolving power $E/\Delta E$ of about 50 judged by performance with the rare gases. This is sufficient to distinguish the doublet components of Ar^+ when using He 584 Å radiation, a difference of 0·18 eV for ~ 5 eV electrons.

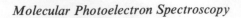

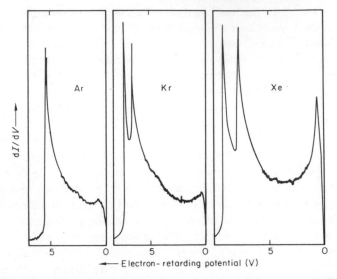

Figure 2.10   Photoelectron spectra for argon, krypton and xenon excited by helium
resonance radiation in the apparatus of Figure 2.1

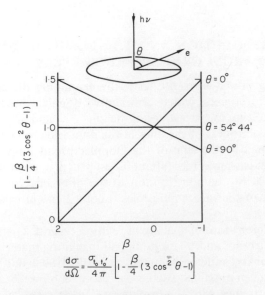

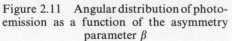

Figure 2.11   Angular distribution of photo-
emission as a function of the asymmetry
parameter β

The peaks are seen to be broadened on the low-energy side and it is this marked asymmetry (cf. Figure 2.10) which is a major limitation to attaining high resolution. It seems to arise partly from electron scatter at the first grid but is partly an inherent defect of using cylindrical grids. Electron photo-emission at low excess energies often follows a $\cos^2 \phi$ law where ϕ is the angle between the electric vector of the incident light and the electron trajectory (Figures 2.11 and 2.12).* Electrons for which $\phi = 0$ enter the retarding field with only a part of their kinetic energy ($E \cos^2\phi$) available for penetration of this field, and are thus rejected at less than the maximum retarding field. In addition, electrons originating at points remote from the grid axis suffer similar loss of effective kinetic energy (Figure 2.13). It can be shown that to avoid peak broadening due to this latter phenomenon, the photon beam must be within 0·5 mm of the axis of a 7 mm diameter grid.

The low energy tail due to the $\cos^2 \phi$ nature of the emission could be eliminated in principle by the use of spherical geometry. McDowell and col-

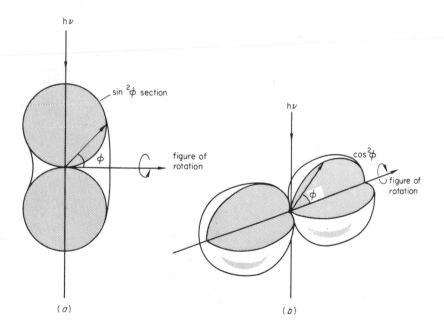

Figure 2.12   Angular distribution of photoelectrons (*a*) $\sin^2 \phi$ (*b*) $\cos^2 \phi$ in form (see text)

---

* There is, however, some evidence that in a few cases the distribution is isotropic, or may even contain $\sin^2\phi$ component.

leagues[13] have obtained considerable improvement in this respect using a spherical grid structure surrounding a photoionization region small compared with the grid diameter (Figure 2.14).

5. DEFLEXION-TYPE PHOTOELECTRON SPECTROMETERS

The use of a focusing deflecting field, either magnetic (180°) or electrostatic (127°), offers the promise of much greater energy resolution and an approach to the ideal spectrum with symmetrical peaks, since electron scatter can be minimized and only those electrons moving parallel to the electric vector need be used, though at the loss of some sensitivity. Both of these spectrometer types have been investigated in the author's laboratory. A small 180° magnetic deflexion analyser (Figure 2.15), having fixed slits and

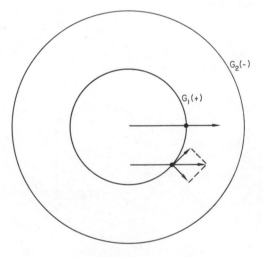

Figure 2.13 Loss of effective kinetic energy for electrons emitted away from the axis of a cylindrical (or spherical) retarding-field analyser

low resolving power, demonstrated that the improvement in peak shape alone greatly extended the usefulness of the method for estimating vertical ionization potentials and Franck–Condon factors in ionization.[14] A larger 127° electrostatic velocity selector with adjustable slits has been described.[12] It is feasible to reach a resolving power $E/\Delta E$ of the order of 1000 with this instrument. If this can be attained, a resolution in the He 584 Å photo-

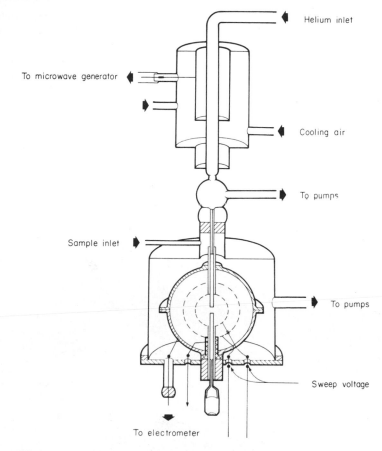

Helium inlet

To microwave generator

Cooling air

To pumps

Sample inlet

To pumps

Sweep voltage

To electrometer

Figure 2.14   Spherical grid photoelectron spectrometer reproduced
from Reference 13 by permission

electron spectra of fine structure with about 5 mV spacing should be possible.
This energy is of the order of some rotational quanta so that a vibration–
rotation analysis for ions may eventually be attainable. The best that has
been achieved at the time of writing is a full-width at half-height of 10 mV
for the peaks in the argon spectrum. Since this is less than kT at room temperature (~ 24 mV), we can obtain some indication of the shape of the
rotational envelope. In a diatomic molecule which undergoes a large
dimensional change on ionization, this reveals the 'shading' to higher or
lower electron kinetic energies, depending on whether the bond length
increases or decreases.

The literature of electron velocity analysis records a number of other forms

of focusing analyses and the reader is referred to Klemperer's reviews where
their various merits are discussed. The two main requirements of high collecting power Ω (the solid angle from which electrons can be collected) and high
resolving power, $E/\Delta E$ where ΔE is the full width of a peak of energy E at half
the maximum height (f.w.h.m.). These two tend to be mutually exclusive and
the choice of analyser for a particular application usually represents some
sacrifice in one or other. The need for high resolving power has dominated
the design of analysers for the X-ray photoelectron field with energies in
excess of 1000 eV for to realize even a modest f.w.h.m. of 1 eV $E/\Delta E \geqslant 1000$

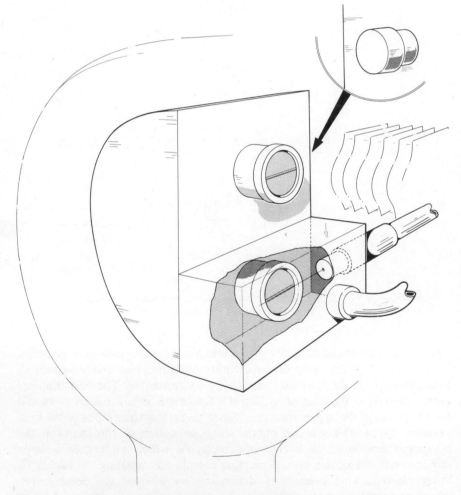

Figure 2.15   The magnetic photoelectron spectrometer of May and Turner (see also
Plates I and II)

PLATE I

(*a*)

(*b*)

Two general views of the magnetic focusing deflexion photoelectron spectrometer described in Reference 2.3 (see also Figure 2.15)

PLATE II

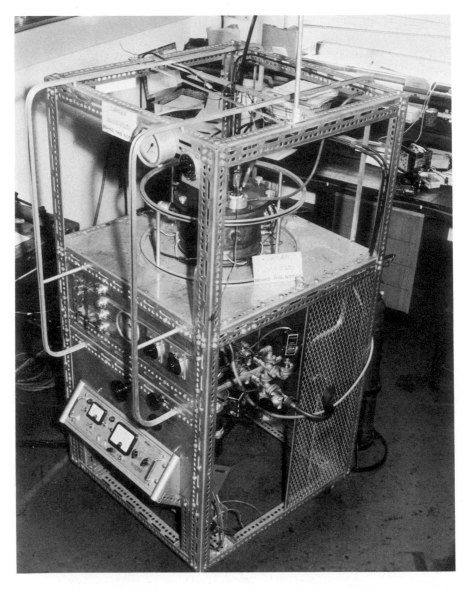

A photoelectron spectrometer for high resolution studies employing electrostatic focusing deflexion (see also Figure 2.16 and Reference 2.12)

PLATE III

(a)   (b)

Target chambers used in the high resolution photoelectron spectrometer shown in Plate II. The target chamber (b) possesses an aperture at the rear, opposite the slit through which the inside slit surfaces could be treated after assembly and in which an accelerator electrode could be inserted

PLATE IV

(*a*)

A commercial high resolution photoelectron spectrometer incorporating a 5 cm radius cylindrical electrostatic field deflexion analyser (reproduced by courtesy of Perkin Elmer Ltd.)

(*b*)

A commercial high resolution photoelectron spectrometer incorporating a 10 cm radius cylindrical electrostatic field deflexion analyser (reproduced by courtesy of Perkin Elmer Ltd.)

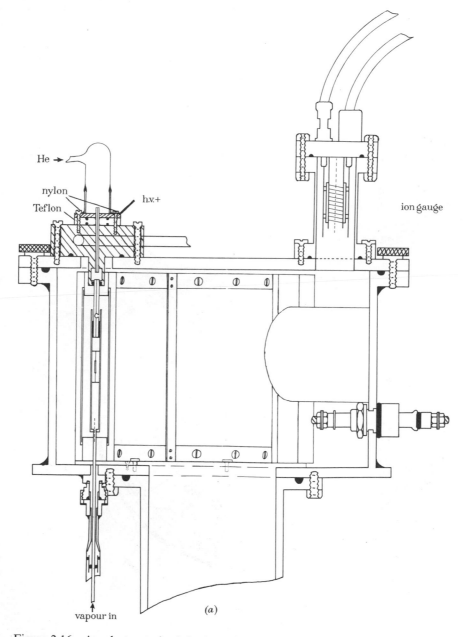

He →

nylon

Teflon

h.v.+

ion gauge

vapour in

(a)

Figure 2.16   An electrostatic deflexion photoelectron spectrometer. (Reproduced from Reference 1.8 by permission). See also Plates III and IV

and this requires much refinement in analyser design and considerable sacrifice in luminosity. Commonly electron counting rates of only a few tens of counts a second have been obtained from vapours in practice.

In contrast where low electron energies of ~ 5 eV are encountered as in the present case a resolving power of 500 leads to f.w.h.m. $= 0.01$ eV and from the relaxation of the focusing precision a correspondingly larger solid angle may be used.

The performance of a given spectrometer can be assessed from the product $\Omega \times E/\Delta E$ but this is an appropriate criterion only for relatively compact sources. Where electrons originating in a large volume are to be analysed some consideration of the ability of an instrument to collect these effectively

Figure 2.16 (*continued*)

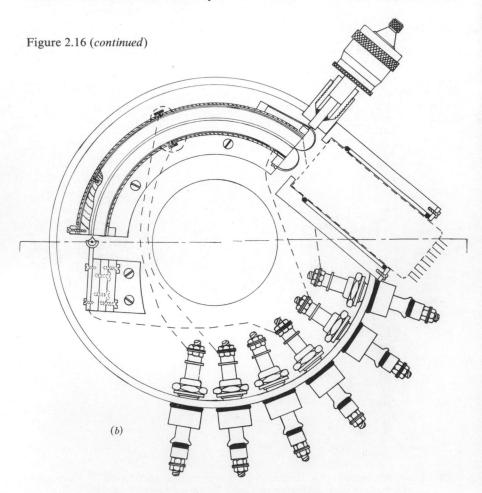

(b)

needs to be given and Geoffrion uses the product of source area and collecting power to define 'luminosity'. This is a relevant consideration in the photoelectron spectroscopy of vapours since at pressures for which electron collision is improbable the light absorption is also small and photoelectrons radiate from a long column of vapour.

Both the spherical deflection analyser described by Purnell and the cylindrical mirror analyser discussed recently by Zashkwara and coworkers, by Hafner and coworkers, and used by Blauth offer a higher figure of merit than most other forms of analyser. From its geometry however the latter is only well suited to nearly point sources and in view of the small 'luminosity' which this implies is not likely to be so useful in photoelectron spectroscopy of vapours as in the study of solids.

REFERENCES

1. M. I. Al-Joboury and D. W. Turner, *J. Chem. Soc.*, **1963**, 5141.
2. F. I. Vilessov, B. C. Kurbatov and A. N. Terenin, *Dokl. Akad. Nauk SSSR*, **138**, 1329 (1961).
3. M. I. Al-Joboury, D. P. May and D. W. Turner, *J. Chem. Soc.*, **1965**, 6350.
4. D. W. Turner and D. P. May, *J. Chem. Phys.*, **46**, 1156 (1967).
5. T. N. Radwan, *Ph.D. Thesis*, London University (1966).
6. A. D. Baker, C. R. Brundle and D. W. Turner, *Intem. J. Mass. Spectr. Ion Phys.*, **1**, 443 (1968).
7. W. C. Price, personal communication.
8. O. Klemperer, *Electron Optics*, 2nd edn., Cambridge University Press, Cambridge, 1953.
9. O. Klemperer, *Rept. Progr. Phys.*, **28**, 77 (1965).
10. R. I. Schoen, *J. Chem. Phys.*, **40**, 1830 (1964).
11. D. W. Turner, *Nature*, **213**, 795 (1967).
12. D. W. Turner, *Proc. Roy. Soc. (London)*, Ser. A, **307**, 15 (1968).
13. D. C. Frost, C. A. M. Dowell and D. A. Vroom, *Proc. Roy. Soc. (London)*, Ser. A, **296**, 568 (1967).
14. D. W. Turner and D. P. May, *J. Chem. Phys.*, **45**, 471 (1966).
15. C. Geoffrion, *Rev. Sci. Instr.*, **20**, 638 (1949).

CHAPTER 3

Atoms and Diatomic Molecules

1. THE INERT GASES ARGON, KRYPTON, XENON

The photoelectron spectra of these gases are shown in Figures 3.1–3.3. They all exhibit two peaks, corresponding to ionizing transitions to the $^2P_{3/2}$ and $^2P_{1/2}$ states of the ion. These ionization potentials are accurately known from spectroscopic data, and so the rare gases are used to calibrate accurately the I.P. scale in all other photoelectron spectra.

The values used here are:

	$^2P_{3/2}$ eV	$^2P_{1/2}$ eV
Ar	15·759	15·937
Kr	14·000	14·665
Xe	12·130	13·436

The ratios of the peak heights read directly from the spectra are:

	$^2P_{3/2}$		$^2P_{1/2}$
Ar	1·90	:	1
Kr	1·87	:	1
Xe	1·90	:	1

After correction for the different electron band-widths (see Chapter 2) these ratios become

	$^2P_{3/2}$		$^2P_{1/2}$
Ar	1·8(5)	:	1
Kr	1·7(0)	:	1
Xe	1·6(3)	:	1

These ratios are to be interpreted as experimental measurements of the ratio $\sigma_{3/2}/\sigma_{1/2}$, the partial photoionization cross-sections for producing ions in their $^2P_{3/2}$ and $^2P_{1/2}$ states at an impacting photon energy of 21·2168 eV. There does not seem to be a direct relation between this and the ratio of the statistical weights of the two states, which is 2:1, but that is obviously the major factor. For a full treatment of the subject of partial cross-sections for the rare gases and comparisons with other experimental data, the reader is referred to a recent paper by Samson and Cairns.[1] In this, measurements of $\sigma_{3/2}$ and $\sigma_{1/2}$ are reported for a range of wavelengths. The results are reproduced in Figure 3.4. It can be seen that the ratio is almost independent of excess energy ($h\nu - I$) except when autoionization occurs. There, over very small wavelength ranges, sudden changes of $\sigma_{3/2}/\sigma_{1/2}$ are seen. This arises from the autoionizing transition which perturbs the continuum being able to connect only with the $^2P_{1/2}$ state. The high cross-section for the Rydberg absorption which initiates the process then leads to an abnormally high ionization cross-section for ionization to the $^2P_{1/2}$ state.

2. DIATOMIC MOLECULES: HYDROGEN, NITROGEN, NITRIC OXIDE, CARBON MONOXIDE, OXYGEN

Introduction

The diatomic molecules have been examined in some detail over the past few years using at first low resolution and then later high resolution photoelectron spectroscopy.[2–8] The spectra of hydrogen and oxygen reproduced here can be found in Reference 5, but these high resolution spectra of nitrogen, carbon monoxide and nitric oxide have not previously been published.

Some work has been done using other than 584 Å radiation.[7–10]

The theoretical calculations of Franck–Condon factors for transitions from a diatomic molecule in its ground electronic state to various vibrational levels of the ion in its several electronic states have been the subject of a number of calculations.[11–13] Halmann and Laulicht[13] have given data for all the diatomic molecules mentioned here and their calculated values are compared to the experimental Franck–Condon factors by drawing lines of length proportional to the calculated figures under each experimental peak. (In doing this a correction has been made for an instrumental factor concerning the different instrumental band-widths at each electron energy.) When peaks are fully resolved the agreement is thus a measure of the variation of the individual ionization cross sections, σ_i, for the vibrational components with excess energy (21·2-I.P.$_{vib}$) above each threshold (I.P.$_{vib}$).

Several molecular orbital calculations, of varying degrees of sophistication, have been carried out on the carbon monoxide and nitrogen molecules, references to which can be found in a paper by P. E. Cade and coworkers[14] together with a complete discussion of the results concerning nitrogen and the difficulties involved in comparing calculated and experimental I.P.'s. The calculated I.P.'s (Koopmans' Theorem Approximation), and experimental vertical I.P.'s are recorded in Table 3.1.

Table 3.1   Calculated and experimental ionization potentials of carbon monoxide and nitrogen

Molecule	Orbital	Observed vertical I.P. (eV)	Calculated I.P. (Koopmans' Theorem)[14]
N_2	$\pi_u 2p$	16·98	17·10
	$\sigma_g 2p$	15·60	17·36
	$\sigma_u 2s$	18·78	20·92
CO	$\sigma 2p$	14·01	15·09
	$\pi 2p$	16·91	17·40
	$\sigma 2s$	19·72	21·87

(1) *Hydrogen* (Figure 3.5)—*Adiabatic I.P. 15·45 eV*

The hydrogen molecule has only one occupied MO, $\sigma_g 1s$, which is strongly bonding. The photoelectron spectrum[2,3] therefore shows only one band consisting of a long series of resolved vibrational components (up to $v' = 15$), whose vibrational spacing (ΔG) has suffered a large decrease compared to that in the ground state of the molecule (see Table 3.2). The series exhibits a fairly rapid convergence towards the dissociation limit of H_2^+, the calculated value of which is marked in Figure 3.5 by an arrow. Figure 3.6, a plot of $\Delta G (v' + \frac{1}{2})$ against v', demonstrates that the potential energy of $H_2^+ \tilde{X}^2 \Sigma_g^+$ is well represented by a Morse function up to at least $v' = 15$. Comparisons with similar plots for $H_2 X^1 \Sigma_g^+$ (ground state)[15] and the $H_2 B^1 \Sigma_u^{+}$[15] are shown also.

(2) *Nitrogen* (Figures 3.7, 3.8 and 3.9)

The photoelectron spectrum of nitrogen exhibits three bands, corresponding to the removal of an electron in turn from the $(\sigma_g 2p)^2$, $(\pi_u 2p)^2$ and $(\sigma_u 2s)^2$ orbitals to leave the ion in the ionic states $\tilde{X}^2 \Sigma_g^+$, $\tilde{A}^2 \Pi_u$ and $\tilde{B}^2 \Sigma_u^+$.

The adiabatic I.P.'s of each band 15·57(9) eV, 16·69(1) eV, and 18·75(9) eV are in good agreement with previous determinations by other techniques.[16]

The first band (Figure 3.8) confirms that the $\sigma_g 2p$ orbital is nearly nonbonding, with the vibrational spacing showing a small decrease compared to the ground molecular-state (see Table 3.2).

The second band (Figure 3.9), resulting from the removal of a strongly bonding $\pi_u 2p$ electron, has at least seven observable vibrational components, and the decrease in the first vibrational spacing (Table 3.2), together with the length of the series confirm the bonding character of the orbital. There is some convergence of the series, though not as marked as in H_2^+, the vibrational spacing between the $v' = 5$ and 6 components being 0.211 eV, the spacing $v' = 0$ to $v' = 1$ being 0.224 eV.

Table 3.2 Ionization potentials, electronic states, and observed vibrational frequencies, for the molecules hydrogen, nitrogen, carbon dioxide, oxygen and nitric oxide

Molecule	Ionic state	Photo-electron band	Adiabatic I.P. (eV)	Observed vibrational frequency (0–1 Vibrational spacing) (cm^{-1})	
				This work	Other techniques
H_2^+	$\tilde{X}^2\Sigma_g^+$	1st	15·45	2260	2235[25]
H_2	$X^1\Sigma_g^+$	—	—	—	4280[25]
N_2^+	$\tilde{X}^2\Sigma_g^+$	1st	15·57(9)	2100	2191[25]
	$\tilde{A}^2\Pi_u$	2nd	16·69(1)	1810	1850[16]
	$\tilde{B}^2\Sigma_u$	3rd	18·75(9)	2340	2397[25]
N_2	$X^1\Sigma_g^+$	—	—	—	2345[25]
CO^+	$\tilde{X}^2\Sigma_g^+$	1st	14·01(8)	2160	2200[25]
	$\tilde{A}^2\Pi_u$	2nd	16·53(6)	1610	1549[25]
	$\tilde{B}^2\Sigma_u$	3rd	19·68(8)	1690	1706[25]
CO	$X^1\Sigma_g^+$	—	—	—	2157[25]
NO^+	$\tilde{X}^1\Sigma^+$	1st	9·26	2260	2340[19]
	$^3\Sigma^+$	2nd	15·65	1200	—
	$^3\Pi$	3rd	16·54	1610	—
	$^3\Delta$	4th	16·84	1200	—
	$3\Sigma^-$	5th	17·55	1200	—
	$^1\Pi$	6th	18·30	1450	1270[30]
	$^1\Sigma^-$	7th	18·39	1100	—
	$^1\Delta$	8th	19·28	1000	—
NO	$X^2\Pi$	—	—	—	1890[25]
O_2^+	$\tilde{X}^2\Pi_g$	1st	12·07(0)	1780	1860[25]
	$\tilde{a}^4\Pi_u$	2nd	16·12	1010	1025[25]
	$\tilde{A}^2\Pi_u$?	3rd?	—	Drawn as 887 (see Text)	887[25]
	$\tilde{b}^4\Sigma_g$	4th	18·7	1090	1180[25]
	$^4\Sigma_u$ or $^2\Sigma_g^-$?	5th	20·29	1130	—
O_2	$X^3\Sigma_g$	—	—	—	1568[25]

As can be seen from the spectrum, the third band arises from the ejection of a virtually nonbonding electron. The electron comes, in fact, from a nominally antibonding orbital $(\sigma_u 2s)^2$, and a very weak antibonding character can be attributed to it as there is a slight increase in the vibrational spacing of the ion compared to that of the molecule.

(3) *Carbon monoxide* (Figures 3.10, 3.11, 3.12 and 3.13)

The spectrum of carbon monoxide is very similar to that of the isoelectronic molecule nitrogen, and an exactly similar description of the three bands, in terms of the orbitals from which electrons are ejected, is possible.

From the spectrum the $\sigma_g 2p$ orbital (1st band, Figure 3.11) appears even less bonding than in nitrogen, while the $\sigma_u 2s$ orbital, though theoretically antibonding, might appear to have some slight bonding character. The $\pi_u 2p$ orbital is strongly bonding, and the photoelectron band (Figure 3.12) relating to the removal of an electron from this orbital shows a similar convergence to that in nitrogen.

The adiabatic I.P.'s are 14·01(8) eV, 16·53(6) eV and 19·68(8) eV, in close agreement with the spectroscopic values.[17] These values are recorded in Table 3.2 as are the observed vibrational frequencies and the frequency values obtained by other techniques.

(4) *Oxygen* (Figure 3.14)

The ground state configuration of oxygen may be written:

$$KK\,(\sigma_g 2s)^2\,(\sigma_u 2s)^2\,(\sigma_g 2p)^2\,(\pi_u 2p)^4\,(\pi_g 2p)^2\;{}^3\Sigma_g{}^-$$

the oxygen molecule thus possessing two unpaired electrons in the $\pi_g 2p$ orbital. This means that a total of nine separate ionic states can be produced by removal of an electron from the valence shell orbitals. These are $O_2{}^+\,{}^2\Pi_g$, formed by removal of a $\pi_g 2p$ electron; $O_2{}^+\,{}^4\Pi_u$, ${}^4\Sigma_g$, ${}^4\Sigma_g$ formed by removal of an electron of antiparallel spin to those in the $\pi_g 2p$ orbital from the remaining valence shell orbitals in turn; and $O_2{}^+\,{}^2\Pi_u$, ${}^2\Sigma_g$, ${}^2\Sigma_u$ and ${}^2\Sigma_g$, formed by the removal of an electron of parallel spin. The doublet states would be expected to have higher energies than their corresponding quartet states (Hund's Rule).

Reviews on the ordering of the ionic states of oxygen deduced by methods other than photoelectron spectroscopy can be found in References 18–20. Only the $\tilde{X}\,{}^2\Pi_g$, $\tilde{a}\,{}^4\Pi_u$, $\tilde{A}\,{}^2\Pi_u$ and $\tilde{b}\,{}^4\Sigma_g{}^-$ levels have been well characterized.

The photoelectron spectrum[5] exhibits at least four bands, possibly five. The adiabatic I.P.'s and observed vibrational frequencies are collected in Table 3.2. The first band corresponds to the $O_2{}^+\,{}^2\Pi_g$ state, and gives an adiabatic I.P. of 12·07 eV in good agreement with previous determinations.[21,22] The increase in vibrational frequency, compared to the ground state of the molecule (see Table 3.2) shows that the π_g2p orbital has some anti-bonding character as expected.

The second band (adiabatic I.P. 16·12 eV) consists of a long series (possibly 20 components) and results from the removal of a strongly bonding $\pi2p$-electron (cf. nitrogen and carbon monoxide) to leave the ion in its $^4\Pi_u$ state. Figure 3.14 shows a plot of ΔG $(v' + \frac{1}{2})$ against v' for this state of the ion, demonstrating the convergence of the series.

The well-known O_2^+ $^2\Pi_u$ state[23] also falls in this region of the spectrum (Vertical I.P. 17·7 eV[23]), but it is not immediately apparent from the spectrum. The relevant portion is shown in Figure 10 on an expanded scale, and the predicted positions for the photoelectron peaks for the $O_2^+ \tilde{A}^2\Pi_u \leftarrow O_2^+ {}^3\Sigma_g^-$ transition have been marked. Some deviation in the ΔG $(V' + \frac{1}{2})$ plot (Figure 3.14) for the $^4\Pi_u$ series occurs near 17·7 eV, and it is possible that this might be due to unresolved $^2\Pi_u$ lines.

The next band, which we will therefore term the fourth band, is formed through the removal of a σ_g2p electron to leave the ion in its $\tilde{b}^4\Sigma_g^-$ state. The adiabatic I.P. 18·17 eV is in close agreement with the spectroscopic value,[23] and the decrease in the vibrational frequency compared to that in the ground state of the molecule (Table 3.2) confirms that the orbital has some bonding character.

The fifth band (shown on an expanded scale in Figure 3.15) has an adiabatic I.P. of 20·29 eV corresponding to the unassigned Rydberg series limit at 20·308 eV.[24] Members of the vibrational series of $v' > 2$ show a marked broadening. The dissociation limit of the $O_2^+ \tilde{b}^4\Sigma_g^-$ state is in this region (20·7 eV[19]), and the broadening could be due to interaction between the two states.

The two most probable assignments for this state of the ion are $^4\Sigma_u^-$[5,2] or $^2\Sigma_g^-$.[19] It seems probable that this state is $^2\Sigma_g^-$ though earlier arguments leading to this assignment have been criticized as inconclusive.[5]

(5) *Nitric oxide* (Figure, 3.16, 3.17, 3.18)

The ground state configuration for nitric oxide may be written

$$KK \, (\sigma_g2s)^2 \, (\sigma_u2s)^2 \, (\sigma_g2p)^2 \, (\pi_u2p)^4 \, (\pi_g2p)^1 \, {}^2\Pi$$

the molecule possessing a single unpaired electron. Removal of an electron from the π_g2p orbital will leave the ion in the singlet state $^1\Sigma^+$, whereas removal of an electron in turn from each of the four other valence shell orbitals can leave both triplet and singlet states of the ion (cf. Oxygen quartet and doublet states). The actual spectroscopic designations of the ionic states which may be formed by removal of an electron are shown below; indicating which electron has been removed.

Molecule $\quad$ $^2\Pi$ KK $(\sigma_g2s)^2$ $(\sigma_u2s)^2$ $(\sigma_g2p)^2$ $\quad(\pi_u2p)^4$ $\quad(\pi_g2p)^1$

Ion $\begin{cases}\text{Triplet} \qquad\qquad {}^3\Pi \qquad {}^3\Pi \qquad {}^3\Pi \quad {}^3\Delta\,{}^3\Sigma^+\,{}^3\Sigma^- \quad — \\[2em] \text{Singlet} \qquad\qquad {}^1\Pi \qquad {}^1\Pi \qquad {}^1\Pi \quad {}^1\Delta\,{}^1\Sigma^+\,{}^1\Sigma^-\;{}^1\Sigma^+\end{cases}$

That there are six states resulting from the removal of a π_u2p-electron and not just one (cf. oxygen where two states can be formed) is due to the presence of the odd electron in the π_g2p-orbital, with an orbital angular momentum, $L = 1$. Thus two electrons for which $L = 1$ are present when a π_u2p-electron is removed, resulting in both Δ (L vectors parallel) and Σ (L vectors anti-parallel) electronic states. The distinction between Σ^+ and Σ^- levels arises from the molecule being heteronuclear. For a complete discussion of the spectroscopic notations for electronic states of diatomic molecules see Herzberg's 'The Electronic Spectra of Diatomic Molecules'.[23]

Assignment of the ground state of the ion as $^1\Sigma^+$ is quite certain, but there is still considerable disagreement over the other experimentally observed ionic levels.[7,8,19,25] The High-Resolution photoelectron spectrum of nitric oxide, together with studies by Lindholm[26] on quantum defects found from studies of the Rydberg series of nitric oxide, seems however to resolve the difficulty.

The electronic assignments, adiabatic I.P.'s and the vibrational frequencies observed from the photoelectron spectrum are recorded in Table 3.2.

The first band in the spectrum (Figure 3.17) is very similar to that in oxygen, resulting from the removal of a π_g2p electron. The increase in vibrational spacing confirms the antibonding nature of the orbital.

Between 15 eV and 20 eV the spectrum is very complex, owing to the overlapping of several bands (Figure 3.18). These bands fall into two natural groups—two sharp bands with little associated fine structure, and a group of bands which each have a long vibrational-series. The sharp bands obviously relate to the removal of a nearly nonbonding electron. They are interpreted as representing the $^3\Pi$ and $^1\Pi$ ionic states formed by removal of a σ_g2p-electron, and are therefore the formal equivalents of the fourth and fifth bands in oxygen (assigning the fifth band as $^2\Sigma_g^-$). The adiabatic I.P.'s are 16·5 eV and 18·30 eV respectively, and the decrease in vibrational frequencies compared to the ground state of the molecule indicate that the orbital has some small bonding character associated with it (cf. oxygen).

The other group of bands equally obviously relates to the removal of a strongly bonding electron (namely from the π_u2p orbital), judging from the long vibrational series and the much larger reductions in vibrational frequencies (Table 3.2). We believe that five of the six theoretical ionic states formed on removal of a $\pi_u 2p$ electron (see page 38) are represented in the photoelectron spectrum, the missing state having an energy greater than 21·2 eV. The reasoning behind the specific assignment of these states is based on the quantum defect work of Lindholm,[26] but the resultant assignments are in disagreement with alternative ones given by Collin and coworkers.[7,8]

Conclusions

The close relation between the four diatomic molecules nitrogen, carbondioxide, nitric oxide and oxygen, is apparent from their photoelectron spectra. In particular the ionic energy-level diagram, shown in Figure 3.19, clearly illustrates the 'aufbau' principle. There is a large gap in energy between the closed-shell nitrogen configuration and the remaining occupied orbital, $\pi_g 2p$ (antibonding), of nitric oxide and oxygen. The $\pi_u 2p$ orbital remains at approximately the same relative energy throughout the series and is strongly bonding throughout. The $\sigma_g 2p$ orbital can be seen to change its character on progressing from carbon monoxide to oxygen, starting as a completely nonbonding orbital and ending up as a moderately bonding one. It should be noted that Figure 3.19 refers to the relative *ionic* energy levels, and is not an accurate molecular orbital energy level diagram owing to the deficiencies of Koopmans' Theorem. The relative ordering of the ionic levels in nitrogen $(\tilde{X}^2\Sigma_g^+, \tilde{A}^2\Pi_u, B^2\Sigma_u^+)$, is a good illustration of this, since the ground state molecular orbital energies are in the order $(\sigma_u 2s)^2, (\sigma_g 2p)^2, (\pi_u 2p)^4$, and the chemistry of the ground state nitrogen molecule is based on this ordering.

3. HALOGEN ACIDS

The ionization of the halogen acids, and of the halogens themselves, has been previously studied by the techniques of electron impact[29–32] and by photoionization.[33] In addition, the photoelectron spectra have been reported, having been obtained by Frost[34], using a spherical grid analyser, and by Price,[35] using a slotted-grid analyser.

The spectra obtained using a 127° electrostatic analyser are reproduced in Figures 3.20, 3.23 and 3.26. The separation in energy of the $^2\Pi_{3/2}$ and $^2\Pi_{1/2}$ states of the ions is clearly resolved in all cases (Figures 3.21, 3.24, 3.26), the observed values being 0·08, 0·33 and 0·66 eV for HCl^+, HBr^+ and HI^+ respectively, in exact agreement with the spectroscopic values.[36,37] This ionic state is also observed to be formed in vibrationally excited levels in all three cases, and the stretching frequencies observed are tabulated (Table 3.3).

The band at high ionization potential relates to the formation of the ions in the $^2\Sigma^+$ states by the loss of an electron from the carbon–halogen σ bond.

Vibrational structure is again observed in all three cases (Figures 3.22, 3.25, 3.26).

Table 3.3   The stretching frequencies observed in the halogen acid ions

	H—X stretching frequency of the ion (cm^{-1})			Molecular frequency (cm^{-1})
	$^2\Pi_{3/2}$	$^2\Pi_{1/2}$	$^2\Sigma^+$	
HCl	2660	2660	1610[a]	2886
HBr	(b)	2420	1290[a]	2560
HI	2020	2100	1300[c]	2230

(a) The separation between the $v = 0$ and $v = 1$ peaks of the $^2\Sigma^+$ band was used to give the vibrational frequency, in view of the convergence of higher members of the series.

(b) The $v = 1$ peak associated with the $^2\Pi_{3/2}$ state falls under the $v = 0$ peak of the $^2\Pi_{1/2}$ state.

(c) This value is subject to a rather large error owing to the breadth of the vibrational peaks.

Comparison of the $^2\Sigma^+$ bands for the three halogen acids shows that in the HI spectrum (Figure 3.26) the vibrational peaks are very broad and indistinct, in the HBr spectrum (Figure 3.25) the first four members are well resolved but higher vibrational levels became broad, and in the HCl spectrum (Figure 3.22), all vibrational peaks are narrow and well resolved. There is a marked convergence of the peaks in the HCl$^+$ $^2\Sigma^+$ band. These phenomena can be interpreted in terms of the varying stability of the $^2\Sigma^+$ ionic states through the series. Price[35] has drawn the potential energy curves for the ground and low ionic states of the halogen acids. The $^2\Sigma^+$ levels are shown as being crossed by a $^4\Pi$ repulsive curve going to fragmentation of the ions to give H(^2S) and X$^+$(^3P). Evidence from the photoelectron spectra suggests that the energy at which the actual curve crossing occurs is high enough in HCl$^+$ to be outside the Franck–Condon envelope, i.e. above $v = 9$. In HBr$^+$, it occurs between $v = 3$ and $v = 4$, and in HI$^+$, it must be below, or in the region of $V = 0$.

In Table 3.3 are listed the vibrational frequencies observed in the $^2\Pi$ and $^2\Sigma^+$ ionic states, and compared with the molecular frequencies. The reduction, upon ionization, in the $^2\Sigma^+$ state is much greater than in the $^2\Pi$ state, owing to the much more highly bonding nature of the orbital.

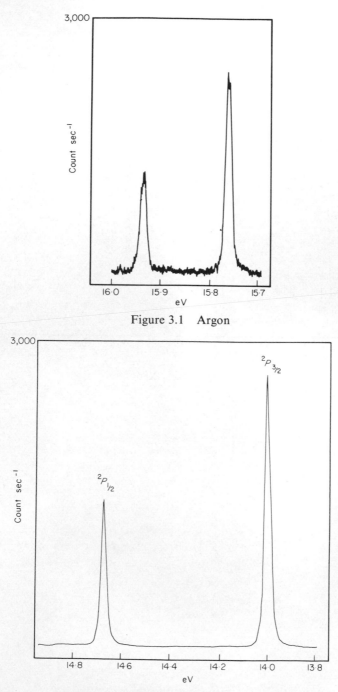

Figure 3.1   Argon

Figure 3.2   Krypton

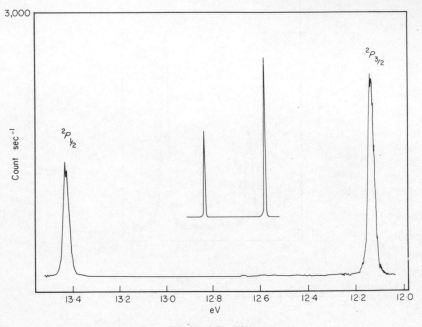

Figure 3.3   Xenon

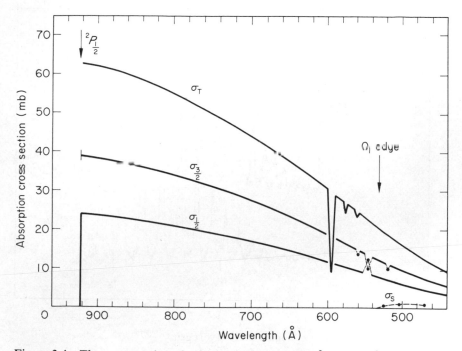

Figure 3.4 The cross-sections for ionization to the $J = \frac{3}{2}$ and $J = \frac{1}{2}$ states of the rare gas ions ($\sigma_{3/2}/\sigma_{1/2}$), after Samson and Cairns, Reference 1. Note that the precise form of the peaks in the $\sigma_{3/2}$ and $\sigma_{1/2}$ curves is uncertain since only single measurements were made at each singularity

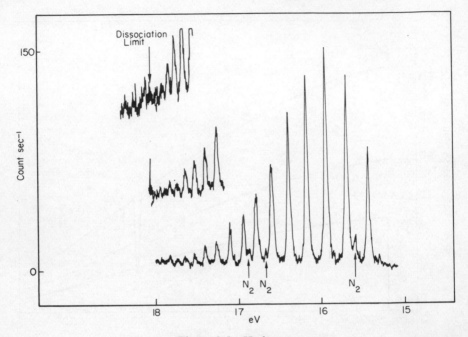

Figure 3.5   Hydrogen

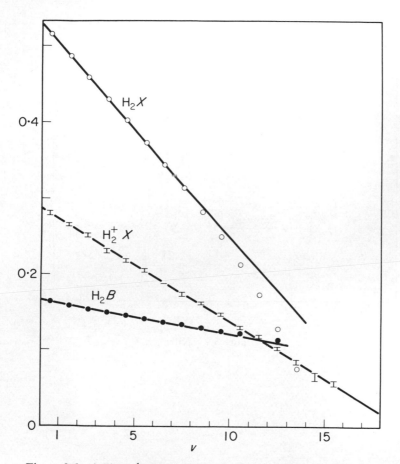

Figure 3.6 $\Delta G/v + \frac{1}{2}$ plots for the vibrational levels in H_2^+ and in two states of H_2 (cf. Reference 2)

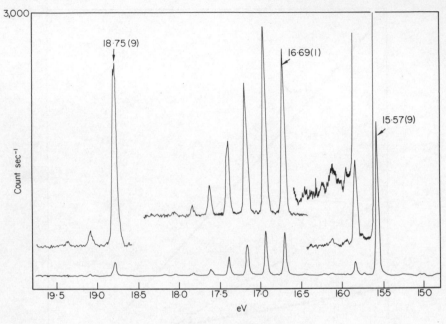

Figure 3.7   Nitrogen

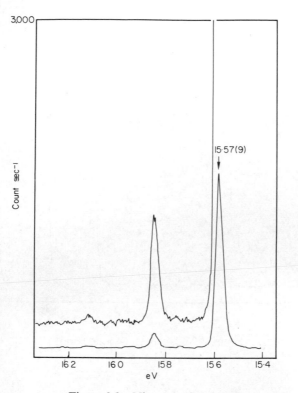

Figure 3.8   Nitrogen (first band)

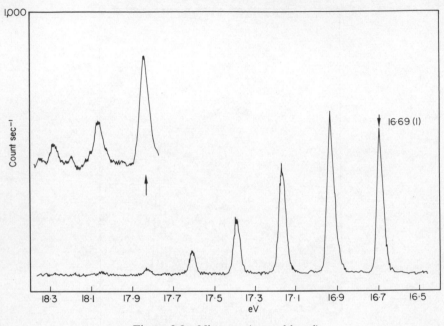

Figure 3.9   Nitrogen (second band)

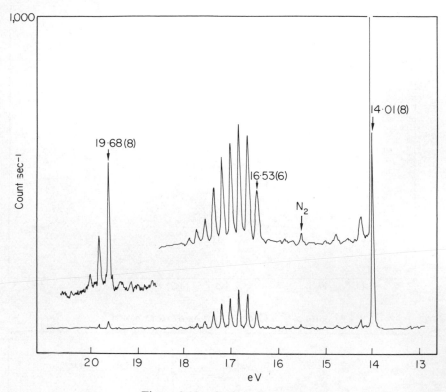

Figure 3.10 Carbon monoxide

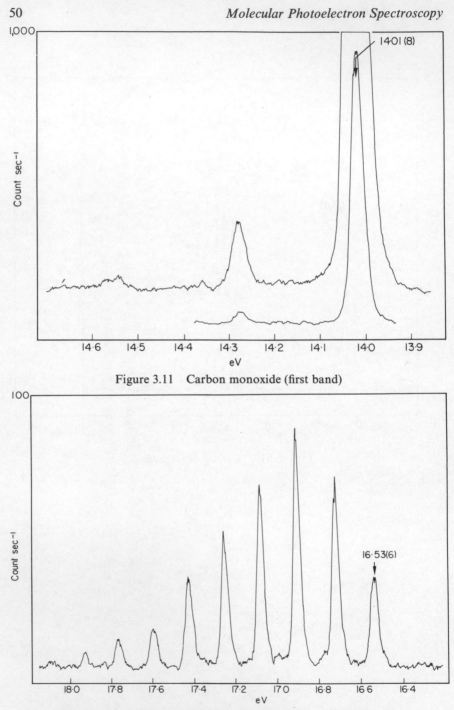

Figure 3.11   Carbon monoxide (first band)

Figure 3.12   Carbon monoxide (second band) to expanded scale

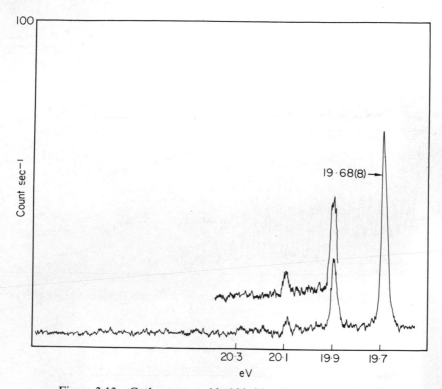

Figure 3.13   Carbon monoxide (third band) to expanded scale

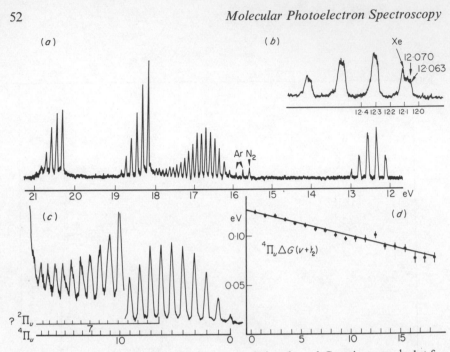

Figure 3.14   Oxygen. Inset, sections to expanded scale and G against $v + \frac{1}{2}$ plot for second band

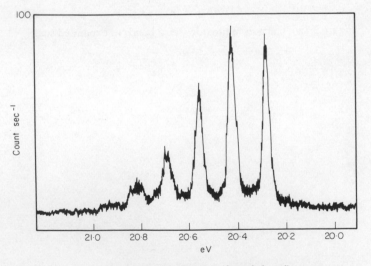

Figure 3.15   Oxygen (fourth band)

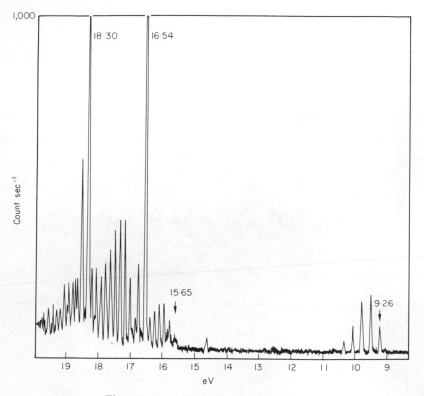

Figure 3.16 Nitric oxide (full spectrum)

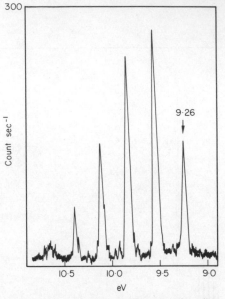

Figure 3.17   Nitric oxide (first band)

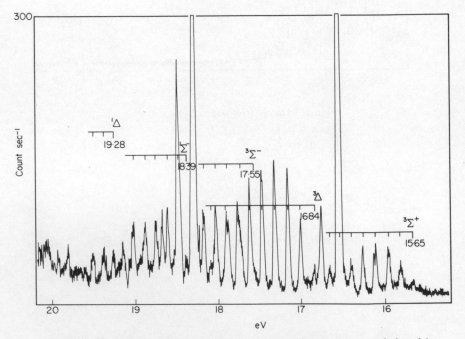

Figure 3.18   Nitric oxide (lower electron energy bands to expanded scale)

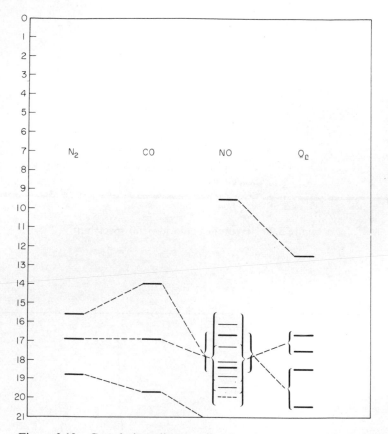

Figure 3.19 Correlation diagram for the photoelectron bands in N_2, CO, NO and O_2 (Vertical Ionization Potentials)

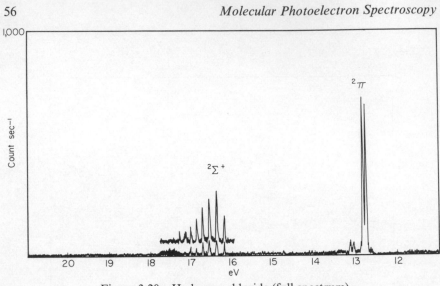

Figure 3.20   Hydrogen chloride (full spectrum)

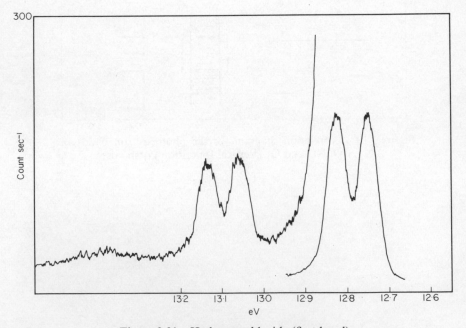

Figure 3.21   Hydrogen chloride (first band)

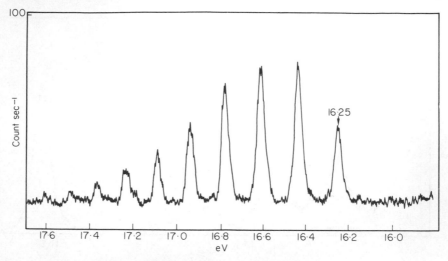

Figure 3.22   Hydrogen chloride (second band)

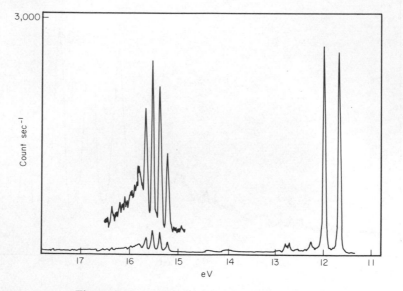

Figure 3.23   Hydrogen bromide (full spectrum)

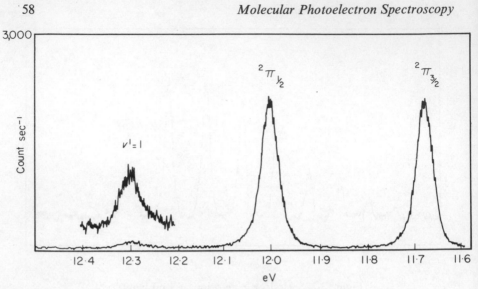

Figure 3.24   Hydrogen bromide (first band)

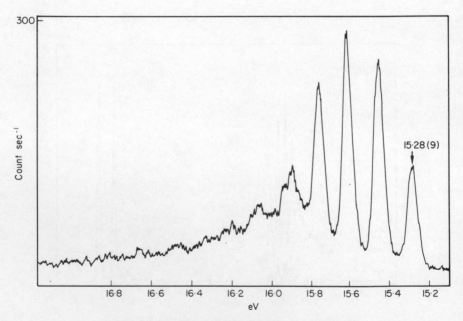

Figure 3.25   Hydrogen bromide (second band)

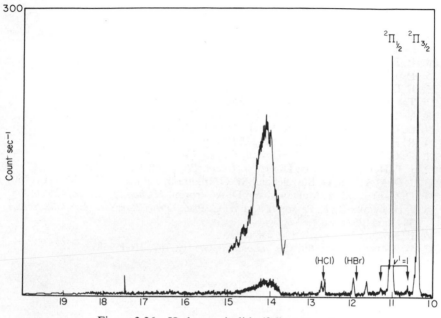

Figure 3.26   Hydrogen iodide (full spectrum)

REFERENCES

1. J. A. R. Samson and R. B. Cairns, *Phys. Rev.*, **173**, 80 (1968).
2. M. I. Al-Joboury and D. W. Turner, *J. Chem. Soc.*, **1963**, 5141.
3. M. I. Al-Joboury, D. P. May, and D. W. Turner, *J. Chem. Soc.*, **1965**, 616.
4. D. W. Turner and D. P. May, *J. Chem. Phys.*, **45**, 471 (1966).
5. D. W. Turner, *Proc. Roy. Soc. (London)*, Ser. *A*, **307**, 15 (1968).
6. D. C. Frost, C. A. McDowell and D. A. Vroom, *Proc. Roy. Soc. (London)*, Ser. *A*, **296**, 566 (1967).
7. J. Collin and P. Natalis, *Chem. Phys. Letters*, **2**, 194 (1968).
8. J. Collin and P. Natalis, *Intern. J. Mass Spectr. Ion Phys.*, In the press.
9. T. N. Radwan, *Ph.D. Thesis*, London University, 1966.
10. W. C. Price, Personal communication.
11. R. W. Nicholls, *J. Res. Nat. Bur. Stand.*, **65**, 451 (1961).
12. M. E. Weeks, *J. Chem. Phys.*, **41**, 930 (1964).
13. M. Hallmann and I. Laulicht, *J. Chem. Phys.*, **43**, 1503 (1965).
14. P. E. Cade, K. D. Sales and A. C. Wahl, *J. Chem. Phys.*, **44**, 1973 (1966).
15. G. Herzberg and L. L. Howe, *Can. J. Phys.*, **37**, 636 (1959).
16. R. E. Worley, *Phys. Rev.*, **89**, 863 (1953).

17. R. E. Huffman, Y. Tanaka and J. C. Larrabee, *J. Chem. Phys.*, **40**, 2261 (1964).
18. K. Codling and R. P. Madden, *J. Chem. Phys.*, **42**, 3935 (1965).
19. F. R. Gilmore, *J. Quant. Spectr. Radiative Transfer*, **5**, 369 (1965).
20. E. Lindholm and H. Sjögren, *Arkiv Fysik*, In the press.
21. K. Watanabe, *J. Chem. Phys.*, **26**, 542 (1957).
22. J. A. R. Samson, *J. Opt. Soc.*, **56**, 769 (1966).
23. G. Herzberg, *Molecular Spectra and Molecular Structure*, Vol. 1, 2nd Edn., Van Nostrand, New York, 1950.
24. R. E. Huffman, Y. Tanaka and J. L. Larrabee, *J. Chem. Phys.*, **40**, 356 (1964).
25. H. Lefebvre-Brion and C. M. Moser, *J. Chem. Phys.*, **44**, 295 (1966).
26. E. Lindholm, Private communication.
27. Taken from *J. Chem. Soc.*, **1965**, Part III. No reference given.
28. Y. Tanaka, *Sci. Papers Inst. Phys. Chem. Res.* (*Tokyo*, **39**, 456 (1942).
29. D. C. Frost and C. A. McDowell, *Can. J. Chem.*, **36**, 39 (1958).
30. R. E. Fox, *J. Chem. Phys.*, **32**, 385 (1960).
31. D. C. Frost and C. A. McDowell, *Can. J. Chem.*, **38**, 407 (1960).
32. J. T. Herron and V. H. Dibeler, *J. Chem. Phys.*, **32**, 1884 (1960).
33. J. D. Morrison, H. Burzeler and M. G. Inghram, *J. Chem. Phys.*, **33**, 821 (1960).
34. D. C. Frost, C. A. McDowell and D. A. Vroom, *J. Chem. Phys.*, **46**, 4255 (1968).
35. J. H. Lemka, T. R. Passmore and W. C. Price, *Proc. Roy Soc.* (*London*), *Ser. A*, **304**, 53 (1968).
36. F. Norling, *Z. Physik*, **104**, 638 (1935).
37. W. C. Price; *Proc. Roy. Soc.* (*London*), *Ser. A*, **167**, 216 (1938).

CHAPTER 4

Triatomic Molecules

1. THE LINEAR TRIATOMIC MOLECULES NITROUS OXIDE, CARBONYL SULPHIDE, CARBON DISULPHIDE AND CARBON DIOXIDE

Introduction

These compounds have been extensively studied using helium 584 Å radiation[1-4], and recently other resonance lines have been used.[5-7] The spectra presented here have been discussed in more detail in Reference 5 which shows how attempts can be made to establish connections with work in fields other than photoelectron spectroscopy, in an effort to obtain a more thorough knowledge of the processes occurring following ionization.

He 584 Å photoelectron spectra and the I.P.'s of the compounds are given here, with an analysis of the vibrational structure of each band. The values of the I.P.'s, and all the experimental vibrational frequencies found for the different ionic states, are collected in Table 4.1, together with the best values found from other techniques for comparison. Experimental Franck–Condon factors (cf. Reference 9, Chapter 1) are given in Table 4.2 together with the few theoretically calculated values.

A full discussion of these results, and their implications, follows on pages 67–76.

(A) Nitrous Oxide, N_2O (Figures 4.1–4.5)

Four Bands can be seen in the spectrum (Figure 4.1), each showing vibrational fine structure.

First band (Figure 4.2)

The adiabatic I.P. (12·89(3) eV) is in good agreement with values found by other techniques.[8,9] The most intense peak ($v' = 0,0,0$) has a width at half-height of approximately 40 mV, which is compatible with a spin–orbit splitting value of 20 mV (cf. References 8, 10).

Three peaks corresponding to vibrationally excited states of the ion can be observed, and from a comparison with the known values for the vibrational frequencies of the N_2O^+ ion in its ground state (Table 4.1) we conclude that v_1 and v_3 are being excited with the frequencies shown in Table 4.1. Carette

Table 4.1   Ionization potential and vibra

Com-pound	Electronic state	Photo-electron band	Adiabatic I.P. (eV)	
			Photoelectron spectroscopy	Other techniques
N_2O^+	$\tilde{X}^2\Pi$	1st	12·89(3)	12·89[8]
	$\tilde{A}^2\Sigma^+$	2nd	16·38(9)	16·39[8]
	$\tilde{B}^2\Pi$	3rd	17·65	—
	$\tilde{C}^2\Sigma^+$	4th	20·11(3)	20·10[8]
N_2O	Ground state of molecule			
COS^+	$\tilde{X}^2\Pi$	1st	11·18(9)	11·18[14]
	$\tilde{A}^2\Pi$	2nd	15·08(0)	—
	$\tilde{B}^2\Sigma^+$	3rd	16·04(2)	16·04[8]
	$\tilde{C}^2\Sigma^+$	4th	17·96(0)	17·93[8]
COS	Ground state of molecule			
CS_2^+	$\tilde{X}^2\Pi_g$	1st	10·60(8)	10·07(6)[8]
	$\tilde{A}^2\Pi_u$	2nd	12·69(4)	—
	$\tilde{B}^2\Sigma_u^+$	3rd	14·47(8)	14·47[8]
	$\tilde{C}^2\Sigma_g^+$	4th	16·19(6)	16·18[8]
	$\tilde{D}^2\Sigma_u^+$	5th	$\simeq 16·6$	—
CS_2	Ground state of molecule			
CO_2^+	$\tilde{X}^2\Pi_g$	1st	13·78(8)	13·76(5)[8]
	$\tilde{A}^2\Pi_u$	2nd	17·32(3)	17·31(2)[23]
	$\tilde{B}^2\Sigma_u^+$	3rd	18·08(2)	18·07(6)[23]
	$\tilde{C}^2\Sigma_g^+$	4th	19·40(0)	19·38[8]
CO_2	Ground state of molecule			

has recently reported structure in electron impact efficiency curves[11,4] cor-. responding to vibrations of the ion in the ground electronic state.

tional frequencies of each state of the ion

Vibrational frequencies (cm^{-1})					
Photoelectron spectroscopy (± 50 cm^{-1})			Other techniques		
$\nu_1^{(a)}$	$\nu_2^{(b)}$	$\nu_3^{(a)}$	$\nu_1^{(a)}$	$\nu_2^{(b)}$	$\nu_3^{(c)}$
1140	—	1750	1126[12]	461[12]	1737[12]
1350	(600?)	2460	1345[12]	614[12]	2451[12]
900?	—	—	—	—	—
1280	—	2300	—	—	—
			1285[12]	589[12]	2224[12]
650	—	2000	610?[14]	—	2069
790	—	2050	—	—	—
—	—	—	—	—	—
970	(410?)	2170	—	—	—
			859[12]	520[12]	2062[12]
—	—	1170	624	205	—
560	—	—	—	—	—
600	—	—	500–600?[8]	—	—
600	—	800?	—	—	—
—	—				
			658[12]	397[12]	1533[12]
1210	—	1420	1250[8]	531[19]	1469[12]
1100	—	—	1131[12]	560[18]	2731[12]
1270	—	1400?	1275[23]	—	—
1390	—	1470	—	—	—
			1388[12]	667[12]	2349[12]

(a) symmetric stretching mode   (b) bending mode   (c) antisymmetric stretching mode.

Second band (Figure 4.3)

The second adiabatic I.P. is 16·38(9) eV. Again an intense 0,0,0 peak is observed, together with well-resolved associated fine structure. Analysis of this structure indicates that v_1 and v_3 are excited with the frequency given in Table 4.1. A small peak at a separation from the 0,0,0 peak which would correspond to a vibration of 600 cm^{-1} has not been identified. Though this value is in agreement with the frequency of v_2[12], the excitation of one quantum of v_2 is forbidden by the selection rules.

This electronically excited state of the ion has been detected by several other techniques.[8,10,11,13]

Third band (Figure 4.4)

The adiabatic I.P. is 17·65(0) eV and the vertical I.P. 18·2 eV. Ionization to this state of the ion, which has not been detected by any other technique, results in a photoelectron band consisting of broad peaks spread over a range of about 1 eV. The vibrational structure is complex and has not yet been successfully analysed, but it almost certainly does not consist of just a simply series in v_1, as was originally thought.[2]

Fourth band (Figure 4.5)

This band (Adiabatic I.P. 20·11(3) eV) exhibits an intense 0,0,0 peak and some weaker components which show that both v_1 and v_3 are excited. The vibrational series cannot consist solely of v_1 components,[2] because the peak at approximately 20·4 eV is not in quite the correct position for a 2,0,0 component, and the intensity pattern would also be irregular if it were the 2,0,0 peak.

(B) Carbonyl Sulphide, COS (Figures 4.6–4.9)

Four bands can be distinguished in the spectrum (Figure 4.6) and the overall form is similar to that of CO_2. Vibrational structure is associated with three of the four bands.

First band (Figure 4.7)

The adiabatic I.P. is 11·18(9) eV. The peaks occur in pairs indicating the splitting of the state into doublet components $^2\Pi_{3/2}$, $^2\Pi_{1/2}$, with a spin–orbit splitting of approximately 44 mV, a value which is in agreement with previous determinations by other techniques.[9,14] Analysis of the associated fine structure indicates that two ion vibrational frequencies are involved. One is v_3,[15] and the other, for which no previous value has been established, is considered to be v_1 by comparison with the molecular ground-state frequency.

Second band (Figure 4.8)

This band (Adiabatic I.P. 15·08(0) eV, Vertical I.P. 15·52(8) eV) consists of a long series of peaks, spreading over approximately 1 eV. This state of the ion has not been observed by any other energy absorption technique, but has been detected in the optical emission spectrum of COS^+.[15] The complex vibrational structure is considered to consist of a series in v_3 and one in v_1, together with combination bands.

Third band (Figure 4.8)

The very intense 0,0,0 peak of this band (Adiabatic I.P. 16·04(2) eV) is overlapped by the tail of the second band. The I.P. is in agreement with previous determinations.[8, 16] No significant vibrational fine structure is observable.

Fourth band (Figure 4.19)

The vibrationally excited components of this band (Adiabatic I.P. 17·9(0) eV) appear with moderate intensity in the photoelectron spectrum, and lead to estimates for the frequencies of v_1 and v_3. The 1,0,1 peak is an uncertain assignment. It is considered that a strong coupling between v_1 and v_3 must be involved, such that the frequency is not correctly represented by the frequency of 1,0,0 plus that of 0,0,1.

In the early work on the photoelectron spectrum of $COS^{1,2}$, it was considered that there might be a fifth band at approximately 20 eV. It has since become clear[4] that peaks observed at approximately 20·3 eV are due to ionization by the hydrogen Lyman β-line emitted in small amounts from impurities in the light source, and are a 'replica' of the first band in the spectrum.

(C) Carbon Disulphide, CS_2 (Figures 4.10–4.15)

The spectrum of carbon disulphide (Figure 4.10) exhibits at least five bands, with the possibility of a sixth (marked Z on Figure 4.10). Vibrational structure is well resolved on four of the bands.

The first band is shown in Figure 4.11. The Adiabatic I.P. is 10·06(8) eV. The spin–orbit components, $^2\Pi_{3/2}$ and $^2\Pi_{1/2}$ of the intense 0,0,0 peak are well resolved, the adiabatic I.P. and the value for the spin–orbit splitting ($\simeq 55$ mV) being in good agreement with previous determinations.[8-10, 15]

A very weak peak at 2340 cm^{-1} (290 mV) separation from the main peak may correspond to the 0, 0, 2 excitation and it is also just possible that there are present excitations involving the bending mode v_2 (see Table 4.2).

The second photoelectron band Figure 4.12, which is interpreted as consisting of a single vibrational series in v_1, spreads over approximately 1 eV

(Adiabatic I.P. 12·69(4) eV, Vertical I.P. 12·83(8) eV). This I.P. has only been observed by photoelectron spectroscopy.[1-4,6] The persistence of a peak at 12·60(7) eV is somewhat puzzling, for although it falls at the first ionization potential of water, it could not be removed by chemical means. It has been proposed[3] as the second adiabatic I.P.

The adiabatic I.P. in the third band (14·47(8) eV)—Figure 4.13— is in agreement with values found by other techniques.[8,13,17] One or possibly two levels of vibrational excitation are observable, yielding a value for the frequency of v_1 for the first time.

There is so far no convincing interpretation for the very weak broad-band centred at approximately 14·1 eV. It cannot represent a direct ionization process. It may of course be due to an impurity or photodecomposition product but we should not exclude the possibility of autoionization of a state of lower energy than 21·2 eV reached by way of, for example, fluorescence after initial absorption of He 584 radiation.

Like the third band, the fourth (Figure 4.14, Adiabatic I.P. 16·19(6) eV) consists of a strong 0,0,0 peak with very little associated fine structure. The vibrational spacings indicate a frequency of 605 cm^{-1}, taken to be v_1, and possibly one of 1650 cm^{-1}. As the selection rules (section III) forbid the excitation of one quantum of v_3, the peak must be the 0,0,2 component (cf. CO_2 and discussion on page 70).

The fifth band (Figure 4.15, Vertical I.P. 17·1 eV) is broad and featureless, spreading over approximately 1·2 eV, with no indication of any resolved fine structure.

As in the case of carbonyl sulphide it at first seemed possible[2] that a further band existed with an I.P. smaller than 21·2 eV. On careful examination under high resolution, two small 'replicas' of the first band were observed at 19·3 eV and 21·0 eV, due to ionization by Lyman β (12·09 eV) and Lyman α (10·20 eV) lines respectively. These must be due to traces of hydrogen-containing impurities entering the light source.

(D) Carbon Dioxide, CO_2 (Figures 4.16–4.19)

The photoelectron spectrum (Figure 4.16) exhibits four bands, each with resolved fine structure. The second and third bands overlap as occurred in carbonyl sulphide, and the spectrum is very similar to that of carbonyl sulphide, with all spectral features displaced some 2 eV to higher ionization energies.

Extensive optical spectroscopic studies of the ionic states of CO_2 have been made over many years.[5,11,18-21]

The first band is shown in Figure 4.17. The width at half-height, 40 mV, of the intense 0,0,0 peak (I.P. 13·78(8) eV) is compatible with a spin–orbit splitting of about 20 mV, in agreement with previous determinations.[8,18]

Two vibrational modes, v_1 and v_3, are excited, their frequencies agreeing with the more recent determinations by other techniques[8,9,22], though not with an earlier quoted value of v_3.[18] The experimental Franck–Condon factors have been discussed recently[4,5] in relation to the theoretical calculations of Sharp and Rosenstock. This is discussed further on page 70.

Carette has resolved fine structure claimed to be due to vibrational excitation on the electron impact efficiency curve for carbon dioxide.[10,4] Only some of this agrees with the vibrational structure in the photoelectron spectrum however. Some of the structure in the electronic impact curve may thus be due to autoionization.

The second band (Figure 4.18, Adiabatic I.P. 17·72(3) eV, Vertical I.P. 17·59(5) eV) consists of a fully resolved simple series of peaks, the ionization potentials and the frequency of the excited mode, v_1, agreeing with those found by other experimental techniques.[18,23]

Associated with the intense $0,0,0$ peak of the third band (Figure 4.18, Adiabatic I.P. 18·08(2) eV) is a peak at 18·23(5) eV which is the $1,0,0$ component. Much weaker peaks to higher energies are at positions such that they may belong to the second or third band series (see Figure 4.18). A very small peak at approximately 18·4 eV possibly represents the $0,0,2$ vibrational level of the third band.

The intense $0,0,0$, peak of the fourth band (Figure 4.19, Adiabatic I.P. 19·40(0) eV) has some very weak vibrational structure associated with it. This is analysed as representing excitation of the v_1 and v_3 modes, and the very large reduction in v_3 compared to that in the ground state is discussed on page 70, together with the similar effect noticed in carbon disulphide.

2. GENERAL DISCUSSION OF THE RESULTS FOR NITROUS OXIDE, CARBONYL SULPHIDE, CARBON DISULPHIDE, AND CARBON DIOXIDE

(A) The Electronic Structure of the Compounds and the Bonding Characteristics of the Orbitals Involved

The electronic assignment of each ionic state of the compounds is indicated in Table 4.1.

The original Mulliken description[24] of the molecular orbital structure of CO_2 was:

$$(1s_O)^2 \ (1s_O)^2 \ (1s_C)^2 \ (s_O + s_C + s_O, \ \sigma_g) \ (s_O - s_O, \ \sigma_u)^2 \ (\sigma_O + \sigma_C + \sigma_O, \ \sigma_g)^2$$

$$O—C—O \qquad O\leftrightarrow O \qquad O—C—O$$

$$(\sigma_O + \sigma_C - \sigma_O, \ \sigma_u)^2 \ (\pi_O + \pi_C + \pi_O, \ \pi_u)^4 \ (\pi_O - \pi_O, \ \pi_g)^4$$

$$O—C—O \qquad O—C—O \qquad O\leftrightarrow O$$

4

the valence shell part of which was simplified to:

$$(\sigma_g)^2 \quad (\sigma_u)^2 \quad (\sigma_g)^2 \quad (\sigma_u)^2 \quad (\pi_u)^4 \quad (\pi_g)^4 \ldots \; {}^1\Sigma_g{}^+$$

o: o: o—c—o   o—c—o   o—c—o   o:

type (*a*)   type (*b*)

Thus the $(\pi_g)^4$ orbital is regarded as being nonbonding ('lone-pairs'), and the (*a*) type $(\sigma_g)^2$ and $(\sigma_u)^2$ orbitals can be written as $(2s_O)^2$, $(2s_O)^2$, 'inner' oxygen atomic orbitals.

Similar configurations may be written for CS_2, COS, and N_2O, and removal of an electron in turn from the π_g, π_u, σ_u, σ_g ... orbitals leaves the ion in the electronic states:

$$\tilde{X}^2\Pi_g, \; \tilde{A}^2\Pi_u, \; \tilde{B}^2\Sigma_u{}^+, \; \tilde{C}^2\Sigma_g{}^+ \ldots$$

each of the states corresponding to an I.P. observed in the photoelectron spectrum.

There are two important questions concerning the Mulliken electronic structure. The first is whether the orbital energy-levels are correctly ordered, and the second is how far can one draw a sharp distinction between orbitals of type (*a*) and (*b*)?

The molecular orbital energy-levels found more recently in three major theoretical calculations[25-27] for carbon dioxide are shown in Table 4.3, together with the experimental I.P.'s found from the photoelectron spectrum. The orbital order found in all cases was

$$(\sigma_g)^2 \; (\sigma_u)^2 \; (\sigma_g)^2 \; (\pi_u)^4 \; (\sigma_u)^2 \; (\pi_g)^4$$

(*a*)   (*b*)   (*b*)

The calculations also indicated that type (*a*) orbitals involved much oxygen p_z—carbon orbital mixing, such as to make them strongly bonding[25,27] and the (*b*) type involved strong *s-p* oxygen mixing which rendered them less bonding.

Now, since Mrozowski[18] identified the $\tilde{A}^2\Pi_u$, $\tilde{X}^2\Pi_g$ emission of $CO_2{}^+$ as having an energy of 3·535 eV, identical to the difference between the first and second I.P.'s of carbon dioxide found from the photoelectron spectrum, there can be no doubt that the second band in the photoelectron spectrum relates to the $^2\Pi_u$ ionic state. That the π_u orbital is strongly bonding is borne out by the shape of the band, a long vibrational series being excited with the frequency of v_1 reduced compared to that in the ground state of the molecule.

Thus, provided Koopman's Theorem (see Chapter 1) holds, the valence shell configuration for carbon dioxide may be written as:

$$(\sigma_g)^2 \; (\sigma_u)^2 \; (\sigma_g)^2 \; (\sigma_u)^2 \; (\pi_u)^4 \; (\pi_g)^4$$

On the other hand we might use the difference between the theoretical eigenvalues and the observed I.P.'s as an indication of the extent to which Koopman's theorem fails when making a comparison between σ and π ionization.

The nonbonding character of the π_g orbital is evident from the photoelectron spectrum since the Franck–Condon factor for the $0,0,0$ vibrational level approaches unity. A similar situation exists for the $\tilde{B}^2\Sigma_u^+$ and $\tilde{C}^2\Sigma_g^+$ states of CO_2^+, indicating that the (b) type $(\sigma_g)^2$ and $(\sigma_u)^2$ orbitals are also very nearly nonbonding (in contradiction to the Mulliken description). This implies that a large amount of s-p oxygen mixing occurs in these orbitals in agreement with the theoretical calculations,[25,27] and that the (a) and (b) type distinction of the σ orbitals in Mulliken's original description is not valid.

The same configuration can be adopted for CS_2 and COS, with the same type of bonding characteristics, to account for the form of the bands in the photoelectron spectra. The ordering of the ionic energy levels is again not in question owing to the positive identification of some of the levels in emission spectra.[15,17]

The fifth band in the photoelectron spectrum of carbon disulphide is interesting and can be accounted for in several different fashions. In keeping with the proposed orbital configuration it could correspond to the removal of an electron from the strongly bonding (σ_u) orbital. The complete lack of fine structure must then be explained by invoking either small unresolvable vibrational spacings, curve crossings of potential surfaces resulting in a short lifetime for this state of the ion $(^1\Sigma_u^+)$, or by considering that the ionization observed in the spectrum is taking place to the level of the continuum in the potential energy surface of the ion (i.e. fragmentation by direct ionization, curve B in Figure 1.3). A further possibility is a transition to a repulsive excited state of an otherwise stable electronic configuration of the ion (curve C in Figure 1.3). This latter type of transition would however be a two photon process, and is therefore improbable.

No theoretical calculations have been carried out as yet on CS_2, but the results of an SCF Molecular Orbital calculation on COS[28] are included in Table 4.3. Again, as can be seen from the table, the calculated ordering of the $(\pi_u)^4$ and $(\sigma_u)^2$ orbitals is reversed compared to the experimental order of the ionic states. The orbital bonding characteristics observed from the spectrum agree well with the theoretical description,[28] the larger bonding character of the π_g orbital here as compared to that found in the other molecules being reflected in the stronger transitions to the vibrationally excited levels of the $^2\Pi_g$ ground state of the ion.

The ordering of the ionic states of N_2O^+, from the photoelectron spectrum, differs from that of the other molecules, the $^2\Sigma^+$ and $^2\Pi$ being reversed.

$$\tilde{X}^2\Pi_i,\ \tilde{A}^2\Sigma^+,\ \tilde{B}^2\Pi,\ C^2\Sigma^+$$

This order is confirmed by examination of the emission spectrum[10] of N_2O^+, where the energy of the $^2\Sigma^+ \rightarrow \tilde{X}^2\Pi_i$ transition equals the difference between second and first I.P.'s, not third and first as in carbon disulphide.[17] The bonding characteristics of the individual orbitals are similar to the corresponding orbitals in the other molecules of this group.

Figure 4.20 is an ionic energy-level correlation diagram for the linear triatomic molecules. Nitrous oxide does not fit well with the general trends observed, but this is not unexpected since the central atom is nitrogen whereas in the other compounds it is carbon. For the three carbon centred compounds, the lowering of the I.P.'s for all orbitals is clearly demonstrated as the larger sulphur atoms replace the oxygen atoms ($3p$ orbitals involved in valence shell molecular orbitals instead of $2p$).

Frequencies of the vibrational mode, v_3, in the $\tilde{X}^2\Pi_g$ and $\tilde{C}^2\Sigma_g^+$ ionic states
It was noted earlier (p. 67) that certain differences between the values found for the vibrational frequency of v_3 in the four compounds are not easily accounted for.

Electronic transitions starting from the zeroth vibrational state are subject to the following well-known selection rules.[20]

(1) The final state may have any number of quanta of a vibrational mode which is symmetric to all symmetry species in the molecule.

(2) Only zero or an even number of quanta of a mode which is antisymmetric to a symmetry species are allowed.

This implies that for carbon disulphide and carbon dioxide v_2 and v_3 can only appear in double quanta, while for N_2O and COS v_3 may appear in single quanta, but v_2 should still appear only in double quanta.

If these rules are adhered to there is a rather striking difference between the changes in frequency of v_3 on ionization of the symmetric molecules CO_2 and CS_2, and the unsymmetric molecules N_2O and COS. Both in the $\tilde{X}^2\Pi_g$ and $\tilde{C}^2\Sigma_g^+$ states of CO_2^+ and CS_2^+ v_3 must have undergone a drastic reduction in frequency compared to its value the ground state of the molecule (Table 4.1). For COS^+ and N_2O^+ the $\tilde{C}^2\Sigma^+$ state actually produces a slight increase in the frequency of v_3, while in the $\tilde{X}^2\Pi$ state a small decrease in v_3 occurs for COS^+ and a moderate one for N_2O^+. The frequency changes in N_2O and COS are not incompatible with mainly nonbonding orbitals, but the drastic reductions for CO_2 and CS_2 are not at all what would be expected. There appears to be no rational explanation for this pronounced difference between the two pairs.

(B) Vibrational Franck–Condon Factors

Sharp and Rosenstock[20] have developed a general technique for calculating Franck–Condon factors for most transitions in polyatomic molecules in the harmonic-oscillator approximation. The input data required are the geometrical dimensions, and the vibrational frequencies of the initial and final states. The technique has been applied to the $\tilde{X}$, $^2\Pi_g$, $CO_2^+ \leftarrow X^1\Sigma_g^+$, CO_2, $\tilde{A}$, $^2\Pi_u$, $CO_2^+ \leftarrow X^1\Sigma_g^+$, CO_2 and $\tilde{X}$, $^2\Pi_g$, $CS_2^+ \leftarrow X$, $^1\Sigma_g^+$, CS_2 ionizing transitions.[20] The calculated values are compared with our experimental results in Table 4.2. Other experimental determinations can be found in References 3, 29–31.

Table 4.2 Vibrational Franck–Condon factors in each state of the ion

Com-pound	Electronic state of ion	Vibrational level 1 2 3	Franck–Condon factors			
			Experimental			Cal-culated
			Photoelectron spectroscopy		Fluor-escence	
			(Present data)	(Ref. 29)	(Ref. 30)	(Ref. 20)
N_2O	$\tilde{X}\,^2\Pi$	000	0·91	1·0		
		100	0·05	—		
		001	0·03	—		
		(200)	0·01	—		
		101	0·01	—		
	$\tilde{A}^2\Sigma^+$	000	0·749	0·82	0·7	
		(010)	(0·002)	—	—	
		100	0·164	0·12	0·2	
		001	0·073	0·06	0·1	
		101	0·008	—	—	
		002	0·004	—	—	
	$\tilde{B}^2\Pi$	000	0·023			
		?	0·057			
		?	0·086			
		?	0·127			
		?	0·124			
		?	0·151			
		?	0·111			
		?	0·095			
		?	0·075			
		?	0·050			
		?	0·036			
		?	0·027			
		?	0·018			
		?	0·011			
	$\tilde{C}^2\Sigma^+$	000	0·78			
		100	0·08			
		001	0·11			
		101	0·02			
		200	0·01			

Table 4.2 (*continued*)

Com-pound	Electronic state of ion	Vibrational level 1 2 3	Franck–Condon factors			
			Experimental			Cal-culated
			Photoelectron spectroscopy		Fluor-escence	
			(Present data)	(Ref. 29)	(Ref. 30)	(Ref. 20)
COS	$\tilde{X}^2\Pi$	000	0·48			
		100	0·28			
		(200)	(0·08)			
		001	0·12			
		101	0·06			
	$\tilde{A}^2\Pi$	000	0·04			
		100	0·08			
		200	0·10			
		001	0·06			
		300	0·07			
		101	0·10			
		400	0·06			
		201	0·10			
		(002)	—			
		301	0·08			
	Overlapping 3rd band	102	0·06			
		401	0·06			
		202	0·08			
	$\tilde{B}^2\Sigma^+$	000	0·95			
	$\tilde{C}^2\Sigma^+$	000	0·48			
		100	0·15			
		001	0·24			
		002	0·09			
		003	0·02			
CS$_2$	$\tilde{X}^2\Pi_g$	000	1·0	—	—	0·9
		(020)	(≤0·05)	—	—	0·1
		(040)	(≤0·025)	—	—	0·02
	$\tilde{A}^2\Pi_u$	000	0·105			
		100	0·216			
		200	0·252			
		300	0·189			
		400	0·124			
		500	0·070			
		600	0·031			
		700	0·012			
	$\tilde{B}^2\Sigma_u^+$	000	0·89			
		100	0·10			
		200	0·01			
	$\tilde{C}^2\Sigma_g^+$	000	0·92			
		100	0·08			
		002	?			

Table 4.2 (*continued*)

Com-pound	Electronic state of ion	Vibrational level 1 2 3	Franck–Condon factors			
			Experimental			Cal-culated
			Photoelectron spectroscopy		Fluor-escence	
			(Present data)	(Ref. 29)	(Ref. 30)	(Ref. 20)
CO_2	$\tilde{X}^2\Pi_g$	000	0·812	0·82	—	0·89
		100	0·149	0·18	—	0·11
		200	0·021	—	—	0·01
		002	0·017	—	—	—
	$\tilde{A}^2\Pi_u$	000	0·082	0·08	0·15	0·09
		100	0·180	0·18	0·24	0·21
		200	0·275	0·20	0·26	0·28
		300	0·180	0·22	0·20	0·22
	Overlapping 3rd band	400	0·125	0·15	0·13	0·12
		500	0·094	0·12	—	0·05
		600	0·043	0·04	—	0·02
		700	0·021	—	—	0·01
	$\tilde{B}^2\Sigma_u{}^+$	000	0·848	0·9		
		100	0·126	0·1		
		(200)	0·017	—		
		(300)	0·010	—		
	$\tilde{C}^2\Sigma_g{}^+$	000	0·91			
		100	0·04			
		002	0·05			

In fact erroneous values for some of the vibrational frequencies were used by Sharp and Rosenstock. The most serious error was the use of 2305 cm^{-1} for v_3 in $\tilde{X}^2\Pi_g CO_2{}^+$. Inghram and coworkers[5] estimated that the use of the correct value for v_3 would increase the Franck–Condon factor of the 0,0,2 excitation ($^2\Pi_g$ state) to a magnitude similar to that of the 2,0,0 excitation. They were unable to identify the former excitation in a photoelectron experiment owing to lack of electron energy resolution which did not allow them to separate the 2,0,0 and 0,0,2 components. In the photoelectron spectrum shown here (Figure 4.17) it can clearly be seen that the peak at 14·1 eV is a double one, and that the two components do in fact have similar intensities. The calculated Franck–Condon factors for the $A^2\Pi_u$ state of $CO_2{}^+$ rest on tentative assignments of v_2 and v_3.[18] It can be seen from Table 4.2 that agreement with experimental values is good, and in particular the prediction that only mode v_1 is excited is supported. For CS_2, the prediction that mode v_2 may be excited in the $\tilde{X}^2\Pi_g$ state must remain unproven as the vibrational spacing between the 0, 2, 0 component from the 0, 0, 0 peak would be too small to be resolved (see Figure 4.11).

Table 4.3   Experimental ionization potentials and calculated orbital energies

Com-pound	Orbital	State of ion	Experimental I.P. (eV)*	Calculated orbital energies			
				Ref. 25	Ref. 26	Ref. 27	Ref. 28
	π_g	$\tilde{X}^2\Pi_g$	13·8	11·5	12·0	14·7	
	π_u	$\tilde{A}^2\Pi_u$	17·6†	19	18·7	20·2	
CO_2	σ_u	$\tilde{B}^2\Sigma_u{}^+$	18·1	17·9	17·1	19·6	
	σ_g	$\tilde{C}^2\Sigma_g{}^+$	19·4	19·5	19·7	21·4	
	σ_u	$\tilde{D}^2\Sigma_u{}^+$	—	~42	39·2	41·1	
	σ_g	$\tilde{E}^2\Sigma_g{}^+$	—	~45	41·1	42·7	
	π	$\tilde{X}^2\Pi_g$	11·2				9·4
	π	$\tilde{A}^2\Pi_u$	15·5†				17·5
COS	σ	$\tilde{B}^2\Sigma_u{}^+$	16·0				15·0
	σ	$\tilde{C}^2\Sigma_g{}^+$	18·0				19·7
	σ	$\tilde{D}^2\Sigma_u{}^+$	—				21·0
	σ	$\tilde{E}^2\Sigma_g{}^+$	—				21·0

* This work, Adiabatic Ionization Potential. † Vertical Ionization Potential.

(C) The Nature of the Fragmentation Processes Occurring

No *direct* information on the fragmentation processes can be obtained from photoelectron spectroscopy other than the energy for the onset of dissociation of an ionic state from vibrational convergence or disappearance of fine structure in a continuum. However from a study of the population of vibrationally excited states and careful correlation with photoionization curves of molecular and fragment ions, and absorption and fluorescence data, considerable insight can be obtained. Dibeler and Walker[9] have obtained photoionization efficiency curves of exceptional quality for all the compounds in the present series.

Nitrous oxide

The ionization threshold of N_2O^+ in the photoionization efficiency curve (Figure 4.21) appears as a sharp step, and no steps corresponding to higher I.P.'s are observable owing to the superposition of much autoionizing structure. For the fragment ion, NO^+, two further thresholds are clearly evident above the Appearance Potential.

Figure 4.22 shows schematically the potential energy diagram for the first three molecular ionic states found from photoelectron spectroscopy. The zeroth vibrational levels are marked and also those which can be populated by direct ionization. No convergence of vibrational levels is observed and only a few levels are populated, indicating that the ionizing dissociation limits are to much higher energies. Also marked are Dibeler and Walker's appearance potentials,[9] thermochemical dissociation limits in the ground molecular state, and some dissociation limits in the ionic state, estimated from known ionization potentials and spectral data.[9]

The striking feature about the diagram is that the A.P.'s are all coincident with thermochemical dissociation limits, within experimental error, except in the case of the first threshold of NO^+ which is 0·84 eV above the limit for the process:

$$N_2O + hv \rightarrow NO^+(\tilde{X}\,^1\Sigma^+) + N(^4S^0)$$

Now direct ionization to states (A') and (B') is not observed in the photoelectron spectra, and direct transitions to state (A') are forbidden by spin conservation rules.[9] Also it is obvious that photoionization cannot produce ion fragmentation by direct transitions to any of the ionic states observed, since there is no population of vibrational levels approaching the dissociation limits. Why then do Dibeler and Walker's results show appearance potentials close to dissociation limits, even of states which cannot be reached by direct single photon ionization? One recognized path of fragmentation is by potential energy surface crossing out of the directly populated electronic states of the ion.[32,33] However no populated levels exist in the region of two of the appearance potentials, 15·01 eV and 17·3 eV. An alternative explanation is that curve crossing out of the populated excited molecular levels into the appropriate ionizing dissociation continua occurs.

Cook and coworkers[31] have recently studied the total ionization and absorption efficiency curves for N_2O, and also the fluorescence from excited states. From their results they deduced that a dissociation continuum is interacting with the ionization continuum at the curve crossing (see Figure 4.23). The dissociation process involved appeared to be

$$N_2O \rightarrow N_2(C^3\Pi_u) + O(^3P) \ldots 12·7 \text{ eV}$$

which gives rise to a peak in the fluorescence spectrum owing to the process

$$N_2(C^3\Pi_u) \rightarrow N_2(B^3\Pi_g)$$

Superimposed on this fluorescence continuum are peaks corresponding to Tanaka's III and IV Rydberg absorption series[8] (leading to the second ionization potential), which indicates that molecules excited to the upper states of these series predissociate into $N_2(C^3\Pi_u + O(^3P))$, and $N_2(C^3\Pi_u)$ then radiate as above. However there is no peak corresponding to the $v' = 0$,

$n = 3$ member of series III, although this band has strong ionization and absorption coefficients. This implies that the potential surfaces of the dissociation state crosses the $n = 3$ Rydberg state of series III between the $v' = 0$ and $v' = 1$ levels. This is indicated in Figure 4.23. Also included are the ground state molecular and ionic ($X^2\Pi$) potential curves. Thus it can be seen that all three excited states may interact. The A.P. or NO^+ at 15·01 eV is coincident with the $v' = 1$ level of the $n = 3$ Rydberg series III. A feasible mechanism for the initial formation of fragment $NO^+(\tilde{X}^1\Sigma^+)$ would therefore seem to be excitation to dissociative state C' followed by preionization to the $NO^+(X^1\Sigma^+) + N(^4S_o)$ ionic level, supplemented by additional transitions to Rydberg level $n = 3$ of series III with curve crossing above the $v' = 0$ level into state C'.

The two higher NO^+ thresholds at 16·53 eV and 17·74 eV are also represented by features in the N_2O dissociation continuum reported by Cook and coworkers.[31] This agreement probably indicates that curve crossings out of dissociation continua similar to those indicated above are occurring. The coincidence of these two thresholds with the second and third ionization potentials of N_2O (see Figure 4.22) indicates that curve crossing out of the populated $A^2\Sigma^+$ and $B^2\Pi$ states of N_2O^+ probably also plays a major part in the fragmentation process.

Carbon disulphide

The photoionization efficiency curve for CS_2^+ is shown in Figure 4.24. Once again the appearance potentials of the fragments are close to calculated ionic dissociation limits, but again fragmentation cannot occur by direct ionization. Figure 4.25 is a potential energy diagram, similar to that for N_2O^+, showing all the relevant features.

The threshold of S^+ is probably due to curve crossing out of the populated levels of $B^2\Sigma_u^+$ CS_2^+ to the ionization continuum of the ground $\tilde{X}^2\Pi_g$ state, plus help from any molecular levels which can cross into this state.

The appearance potential of CS^+ is nearly coincident with the fourth I.P. ($\tilde{C}^2\Sigma_g^+CS_2^+$). It has a very sharp onset, suggesting that CS^+ is formed by curve crossing out of the populated $C^2\Sigma_g^+$ state.

It has been shown above how fragmentation by direct ionization to the dissociation limits of the ionic states is always impossible for N_2O and CS_2 (with the possible exception of the fifth band in CS_2) and yet how fragments always occur near recognizable dissociation limits. This is also true for CO_2 and COS, a full discussion of which is given in Reference 4. It must be concluded therefore that for these molecules at least, there exists a sufficient density of accessible potential surfaces such that curve crossing to the ground ionic state at or near the dissociation limit is always possible, and that often a similar situation occurs near higher dissociation limits.

3. BENT TRIATOMIC MOLECULES

(A) Water and Deuterium Oxide

The photoelectron spectrum of water[34-36] (Figures 4.26–4.32) exhibits three bands, indicating three orbital-levels of energy greater than -21.2 eV, and not four as has been reported from electron-impact studies.[37-39] Electron-impact studies have, in facto, provided an unambiguous value only for the first ionization potential. Deuterium oxide gives a spectrum whose overall form is very similar to that of H_2O, but which has the expected differences in the frequencies of the ionic vibrational modes that are excited.

From their MO calculations on the water molecule, Ellison and Shull[34] found the ground-state configuration and orbital energies to be:

$$(1a_1)^2 (2a_1)^2 (1b_2)^2 (3a_1)^2 (1b_1)^2 \ldots {}^1A_1$$

$$\simeq (1s_0)^2 (2s_0)^2 (1b_2)^2 (3a_1)^2 (2px_0)^2$$

Orbital Energies (eV) -577, -36, -18.55, -13.2, -11.79.

The geometrical forms of the molecular orbitals $\psi(3a_1)$ and $\psi(1b_2)$ are illustrated schematically in Figure 4.33, from which it can be seen that $\psi(3a_1)$ may be described as involving H—H bonding character and $\psi(1b_2)$ H—H antibonding character.

The individual bands in the photoelectron spectrum of H_2O and D_2O are recorded and discussed below, and the experimental vibrational frequencies found, collected in Table 44, together with the ground-state molecular-vibrational frequencies.

Table 4.4   Vibrational frequencies of H_2O^+ and D_2O^+ in their different electronic states

Electronic state	Molecule	Photoelectron band	Vibrational frequencies (cm^{-1})	
			$v_1{}^{(a)}$	$v_2{}^{(b)}$
Molecular ground state—${}^1A_1{}^{12}$	H_2O		3652	1595
	D_2O		2666	1178
${}^1B_1$	H_2O^+	1st	3200 ± 50	1380 ± 50
	D_2O^+		2310 ± 50	980 ± 50
${}^2A_1$	H_2O^+	2nd	—	975 ± 50
	D_2O^+		—	715 ± 50
${}^2B_2$	H_2O^+	3rd	2990 ± 100	1610 ± 100
	D_2O^+		2170 ± 100	1210 ± 100

(*a*) symmetric stretching mode   (*b*) bending mode.

First band (Figures 4.27 and 4.30). The Adiabatic I.P. (H_2O = 12·61(6) eV D_2O = 12·62(4) eV) is in close agreement with the spectroscopic value.[40] Several vibrational levels are discernible in the spectrum, analysis of which leads to the frequencies given in Table 4.4. The similarity of these frequencies to those of the ground molecular state demonstrates clearly that we are concerned here with the removal of an essentially nonbonding electron (oxygen lone pair, $(1b_2)^2$) so that little resultant change in molecular dimensions takes place.

$$H_2O^+ \ldots (1b_1)^1, {}^2B_1 \leftarrow H_2O \ldots (1b_1)^2, {}^1A_1$$

The relatively intense 0,0,0 peak supports this. The experimental Franck–Condon factors for the transitions to the different vibrational levels are recorded in Table 4.5.

Table 4.5   Experimental Franck–Condon factors of the ${}^2B_1$ ionic levels of H_2O^+ and D_2O^+

Vibrational component

$v_1{}^{(a)}$	$v_2{}^{(b)}$	$v_3{}^{(c)}$	H_2O^+	D_2O^+
0	0	0	0·757 ± 0·005	0·702 ± 0·005
0	0	0	0·069 ± 0·005	0·087 ± 0·005
1	0	0	0·143 ± 0·005	0·148 ± 0·005
1	1	0	0·013 ± 0·002	0·034 ± 0·002
2	0	0	0·018 ± 0·002	0·025 ± 0·002
2	1	0	⩽0·002	0·004 ± 0·002
3	0	0	⩽0·008	⩽0·002
3	1	0	—	⩽0·002

(a) symmetric stretching mode   (b) bending mode   (c) antisymmetric stretching mode.

Botter and Rosenstock[41] have calculated the Franck–Condon factors for H_2O^+ and D_2O^+ in the ${}^2B_1$ ionic state as a function of both bond angle and distance. The data required for this calculation are the vibrational frequencies of the excited modes. Not having these available at the time they used the frequencies of the $\tilde{C}$, ${}^2B_1$ Rydberg molecular state[12] which are in fact very similar to the ionic ground-state frequencies. Use of the frequencies found from the photoelectron spectrum does not significantly alter their conclusions. By fitting the experimental Franck–Condon factors to their calculated curves of Franck–Condon factor versus bond angle and bond distance, it is possible to obtain an estimate for the bond angle and distance. The values obtained are indicated in Table 4.6, together with the molecular ground-state parameters[12] and the parameters of the C^1B_1 Rydberg state deduced from a rotational analysis.[42] The results are in disagreement with those of

Table 4.6   Geometry of H_2O and D_2O

Electronic state	H_2O		D_2O	
	$r_{eq}(\text{Å})$	$\theta_{eq}(°)$	$r_{eq}(\text{Å})$	$\theta_{eq}(°)$
$^1A_1$-Ground molecular[12]	0·956	105·2	0·957	104·9
C^1B_1-Rydberg[42]	1·016	106·9	1·009	107·6
$^1B_1$-Ground ionic*	0·999	110·3	0·996	110·0

* From the experimental Franck–Condon factors of the photoelectron spectrum—see text.

Krauss,[43] who obtained a bond angle of $119°$ for the H_2O^+ ion in this state using an SCF Gaussian Basis method of calculation. Such a large increase in bond angle is, of course, not at all what would be expected on removal of an essentially nonbonding electron.

Very recently Botter and Rosenstock[44] have extended their calculations to include the effects of anharmonicity upon the calculated Franck–Condon factors, and it has become obvious from these results that too great a precision was expected from the original calculations. A reasonable estimate for the geometry of both ions would now be:

$$r = 0·995 \pm 0·005$$
$$\theta = 109°$$

Improved estimates will rest upon improved knowledge of the anharmonicity bending vibration, v_2.

The second band (Figures 4.28, 4.31, Adiabatic I.P. 13·7 eV) consists of a long series of what at first sight appear as equally spaced peaks, the mean separations of which (0·120 eV (H_2O), 0·089 eV (D_2O)) correspond to the vibrational frequencies given in Table 4.4. Comparison with the molecular ground state frequencies shows v_2, the bending mode, to be the only reasonable assignment here.

The length of the series and the reduction in magnitude of v_2 compared to that observed in the ground state of the molecule indicates that an electron has been removed from a strongly bonding orbital. Now since only the bending mode is excited the major dimensional change on ionization must be in bond angle, and this is consistent with the removal of an electron from an orbital possessing strong H—H bonding character, i.e. the $\psi(3a_1)$ orbital, to leave the ion in its $^2A_1$ electronic state.

Comparison with the isoelectronic species $\cdot NH_2$ is rewarding. The ultraviolet absorption spectrum corresponding to the transition:

$$NH_2(\tilde{A}) \longleftarrow NH_2(\tilde{X})$$
$$^2A_1 \qquad\qquad ^2B_1$$

has been examined in detail by Dressler and Ramsey,[45], and Dixon[46] has shown that the most stable conformation of NH_2 in its upper, $^2A_1$, state is slightly bent (Figure 4.34). The non-linearity causes considerable complexity in the region of the first few vibrational levels of the upper state. The H_2O^+ ion being isoelectronic with the $\cdot NH_2$ radical one might expect a similarity in geometry of the first excited $^2A_1$ state of H_2O^+, with perhaps a similar effect to be found in the first few vibrational levels. On looking carefully at the second band of the photoelectron spectrum (Figure 4.28) it is apparent from comparison with D_2O (Figure 4.31) that the zeroth vibrational position must be rather indeterminate, an apparent 'smearing-out' of the first few vibrational levels occurring. From this it may be concluded that the $^2A_1$ state of H_2O^+ is probably slightly non-linear in a similar fashion to that in NH_2. The effect is an example of a strong 'Renner-splitting' for a more detailed treatment of which the reader is referred to References 45 and 46 for a discussion relating to NH_2, Reference 34 for a more detailed treatment of H_2O, and the general discussion by Herzberg.[12]

The first clearly measurable peaks in the photoelectron series are at 13·89(7) eV for H_2O and 13·98(8) eV for D_2O, and the two series can be extrapolated back to a common origin near 13·7 eV. No peaks can be found associated with this series at a lower I.P. (traces of CO_2 at 13·78 eV) and so 13·7 eV is taken to be the adiabatic I.P. of both compounds. This may be compared to a figure of 14·35 eV found by electron impact.[38]

The third band (Figures 4.29 and 4.32, Adiabatic I.P. H_2O 17·22 eV, D_2O 17·26 eV) is more complex than the other two, consisting of peaks exhibiting varying degrees of broadening. Broadening can indicate an unstable ionic state having a short lifetime. The mass spectrometric appearance potential of OH^+ is 18·01 eV,[39,47] but no upward break is observed in the ionization efficiency curve for H_2O^+ at this value, which indicates that the OH^+ ions are being formed directly when a $1b_2$ electron is removed at and above this energy. It has been suggested[33] that the $^2B_2$ state of the H_2O^+ ion is completely predissociated by curve crossing with the $4A''$ repulsive state to give $OH^+_\circ$. This is illustrated in the potential energy diagram of Figure 4.35.

The third band of the photoelectron spectrum supports this hypothesis in that the first few peaks in the band are somewhat sharper than the rest, only these vibrational levels apparently being below the crossing point with the $4A''$ state.

Vibrational analysis of the band is made difficult by its complexity and supposedly dissociative broadening. A possible series of doublets can be picked out corresponding to the frequencies given in Table 4.4. The reduction in frequencies for D_2O^+ compared to H_2O^+ is about the expected value for replacement of H's by D's. The decrease in the frequency of v_1 and the slight increase for that of v_2, compared to the ground-state molecular frequencies, are consistent with the removal of an electron from an orbital having O—H bonding character, and H—H antibonding character, as is implied in the $\psi(1b_2)$ orbital of H_2O.

Complete certainty as to the assignment of the adiabatic I.P.'s is again difficult since there may be one or two vibrational levels of such low intensity as to have been undetected.

(B) Hydrogen Sulphide, H_2S

The high resolution spectrum of H_2S has not been published previously. The ionization potentials have been deduced from a low resolution spectrum (cylindrical-grid analyser) by Al-Joboury and Turner.[49]

The high resolution spectrum (Figures 4.36–4.39) is similar in gross structure to that of its analogue, H_2O, in that it exhibits three bands, indicating three orbital levels of energy greater than -21.2 eV, in agreement with recent MO calculations.[50-52] The earlier low resolution spectra[49,53] indicated two additional bands at high I.P. (18.0 eV and 20.1 eV), but no trace of these can be found in the high resolution spectrum, and so they may be regarded as due to impurities. The values of the experimental vibrational frequencies, are listed in Table 4.7, together with those of the ground molecular-state.

Table 4.7   Vibrational frequencies of H_2S^+ in its different electronic states

Electronic state	Photoelectron band	Vibrational frequencies (cm^{-1})	
		$v_1{}^{(a)}$	$v_2{}^{(b)}$
Molecular ground state—$^1A_1{}^{12}$	—	2615	1183
$^1B_1$	1st	2380 $\pm$ 50	—
$^2A_1$	2nd	2040 $\pm$ 50?	940 $\pm$ 50
$^2B_2$	3rd	1900 $\pm$ 100?	1470 $\pm$ 100?

(a) symmetric stretching mode   (b) bending mode.

The molecular-orbital structure of H_2S is basically similar to that of H_2O, and the ground-state configuration can be represented as:

$$H_2S \ (1a_1)^2 \ (2a_1)^2 \ (1b_2)^2 \ (3a_1)^2 \ (1b)^2 \ \ldots \ {}^1A_1$$

The composition of the three highest energy orbitals differs from H_2O only in that $\psi(3a_1)$ and $\psi(1b_2)$ involve slight amounts of $3d$, x_2-y^2, z^2 and yz atomic orbitals.[52] Three numerical calculations on H_2S have been carried out,[50–52] and the orbital energies obtained are listed in Table 4.8 together with the

Table 4.8   Molecular orbital energy levels of H_2S

Orbital	Energy (eV)			Experimental vertical I.P.'s (eV)
	Ref. 50	Ref. 51	Ref. 52	
$^1b_1$	− 9·53	− 10·56	− 10·47	10·48
$^2a_1$	− 12·35	− 12·72	− 14·73	13·25
$^1b_2$	− 14·43	− 14·53	− 18·37	15·35
$^1a_1$	− 25·57	− 22·8	− 38·31	—

experimentally found I.P.'s. It is worth noting that all three suggest only three orbitals of energy above $- 21\cdot2$ eV, and that the most recent places the fourth orbital at $- 38$ eV.

From the shape of the first band (Figure 4.31, Adiabatic I.P. 10·48(4) eV) it is obvious that the $1b_1$ electron removed is even less bonding than is H_2O, and the orbital may be regarded as completely $(3p)^2$, a sulphur 'lone-pair'. Only one vibrational mode, v_1, is excited, with very low probability, in agreement with Franck–Condon considerations.[54] The I.P. is in good agreement with the spectroscopic value of 10·47 eV[55] but is somewhat higher than the latest photoionization value,[56] and the earlier photoelectron value quoted by Al-Joboury and Turner, both of which are 10·42 eV. The difference may be due to the broadness of the zeroth vibrational level of this state of the ion.

The second band (Figure 4.38, Adiabatic I.P. 12·76(2) eV, Vertical I.P. 13·25(4) eV) obviously relates to the removal of a strongly bonding electron, as in the case of H_2O, but the fine structure of the band is markedly different. The low I.P. side of the band shows clear fine structure while above the vertical I.P. the structure becomes irregular and confused. The vibrational structure may be analysed in one of two fashions. Either it consists of a simple series in v_2 and the irregularities of the individual peaks are due to a Renner–Teller effect similar to that in this state of H_2O (see page 80), or there is a second series present involving the addition of one quantum of v_1 to each member of the first series. The band has not been examined in nearly such great detail as for H_2O and we must admit the slight possibility of one or two early vibrational members being missed owing to low intensity.

Previously suggested I.P.'s for the removal of an electron from this orbital have been 12·2 eV[37] and 12·46 eV[57], from electron-impact work, and 12·62 from photon-impact work.[56]

The appearance potentials of the fragment ions S^+ and SH^+ have recently been accurately measured by Dibeler and Liston[56] using photon-impact mass spectrometry. The appearance potential of the well-known metastable S^+ peak observed in the mass spectrum is 13·36 eV, and that of 'normal' S^+ 13·40 eV. These values coincide with the energy at which the vibrational series of the second band in the photoelectron spectrum 'breaks off', which suggests that the breaking off is due to a lifetime limitation imposed by curve crossing transitions in this region. Thus ionizing transitions to this state of the ion ($^2A_1$) produce stable H_2S^+ below the crossing region and S^+ fragments in the region. The production of stable H_2S^+ in the $^2A_1$ state is verified by the sharp upward break in the H_2S^+ photoionization efficiency curve[56] at approximately 12·6 eV. The curve reaches a maximum near the appearance potential of S^+ and then falls off as expected. The appearance of $S^+(m)$ could come from the small part of the surface in which the crossing rate is experimentally observable[54] but Dibeler and Liston[56] suggest that because of the different shapes of the efficiency curves of S^+ and $S^+(m)$, they must be regarded as arising from entirely different electronic states. It is difficult to see how S^+ fragments can come from different *ionic* levels, since only the $^2A_1$ ionic level is populated in the 13·4–14·6 eV energy range (see Figure 4.38), the region where most of the differences in structure between the ionization efficiency curves of S^+ and $S^+(m)$ occur. There is some structure on the $S^+(m)$ curve in this region, which could indicate that some of the $S^+(m)$ comes originally from autoionizing levels. We consider that for the main part S^+ and $S^+(m)$ have the same origin, the difference in A.P. being a measure of the Kinetic Shift involved. An accurate study of the kinetic energies of the metastable ions would be of interest in helping to decide their origin.

Fiquet-Fayard and Guyon[33] have constructed a correlation diagram for the energy levels of H_2S^+ and the fragments produced. It is reproduced in Figure 4.40. They have shown that from the correlation rules, there must be curve crossing of a quartet state through both the first excited state ($^2A_1$) and the ground state to yield S^+, and since only the $^2A_1$ state is populated in the region of the curve crossing, all the S^+ must come from this crossing.

The third band of H_2S (Figure 4.39, Adiabatic I.P. 14·56 eV, Vertical I.P. 15·35 eV) is similar to that in H_2O, and there is the same difficulty in establishing the adiabatic I.P. We have placed it at 14·56 eV somewhat arbitrarily. Our value may be compared to those of 14·0 and 14·2 eV obtained by electron impact.[37,57] We think that these values are too low, and that the initial break in the electron-impact efficiency curve is possibly caused by an autoionizing process. There is certainly no evidence of an increase in total ion-current at these values in the photoionization efficiency curves.[56]

The vibrational analysis of the band is uncertain, but it may be treated in a similar fashion to that of H_2O, as two series involving v_1 and v_2.

As can be seen, the band differs from that of H_2O in that *all* the vibrational levels are very broad. Fiquet-Fayard and Guyon[33] considered that SH^+ is formed in an analogous manner to OH^+, i.e. the $^2B_2$ level is predissociated by curve crossing of the $4A''$ state (see Figure 4.40). In fact the A.P. of SH^+ is lower than the adiabatic I.P. of the $^2B_2 H_2 S^+$ state, which explains why all its vibrational components are broadened, but also indicates that formation Of SH^+ below 14·56 eV cannot originate from the $^2B_2$ state. SH^+ may come from curve crossing with the $^2A_1$ ionic state, which is weakly populated vibrationally right up to the start of the $^2B_2$ state. There may also be some contribution from populated autoionizing levels in this region.

Fiquet-Fayard and Guyon have given reasons why the formation of S^+ by curve crossing out of the $^2B_2$ state may be considered to be negligible.

(C) Sulphur Dioxide

Sulphur dioxide possesses 18 valence shell electrons, two more than the stable linear triatomic molecules CO_2, CS_2, N_2O and COS. The electronic structure of these compounds (see also page 68) is:

$$\ldots (\sigma_g)^2 \, (\sigma_u)^2 \, (\pi_u)^4 \, (\pi_g)^4 \, \ldots \, {}^1\Sigma_g{}^+$$

Nonbonding Bonding Nonbonding

The two extra electrons of SO_2 enter what was for the linear compounds the lowest unoccupied orbital (π_u, antibonding). The energy of this orbital decreases rapidly with decreasing bond angle, going over to an a_1 orbital in the bent form. Thus the SO_2 molecule should have a lower energy in the bent form, as is found experimentally ($\theta = 119°$). The bending removes the degeneracy in the $(\pi_g)^4$ and $(\pi_u)^4$ orbital levels which divide to (a_2), (b_2), and (b_1), (a_1) orbitals respectively in the bent form. Thus the ground state molecular configuration may now be represented as:

$$\ldots (3a_1)^2 \, (3b_2)^2 \, (1b_1)^2 \, (5a_1)^2 \, (4b_2)^2 \, (1a_2)^2 \, (6a_1)^2 \, \ldots \, {}^1A_1$$

The shapes of these orbitals are shown schematically in Figure 4.41 and the correlation diagram of the observed vertical I.P.'s of SO_2 and CO_2 is shown in Figure 4.42. The orbitals to which it is considered the I.P.'s relate, and their approximate bonding characteristics are also indicated.

There seems to be little agreement in the literature as to the calculated ordering of the molecular orbitals,[59–61] except that it seems to be generally accepted that $6a_1$ is the highest filled orbital. The calculated value for the first I.P. is given as 10·82 eV,[60] considerably lower than the experimental value of 12·5 eV.

As expected the photoelectron spectrum (Figures 4.43–4.46) of SO_2 is more complex than that of the linear triatomic molecules. A high-resolution spectrum has recently been published elsewhere by Eland and Danby,[62] and our interpretation of the spectrum agrees in general with theirs, with a few minor differences. The experimental I.P.'s, and observed vibrational frequencies are given in Table 4.9, together with the ground state molecular values.

Table 4.9   Ionization potentials and observed vibrational frequencies for the ionic states of SO_2

Orbital	Adiabatic ionization potential (eV)	Vibrational frequency (cm^{-1})		
		$v_1{}^{(a)}$	$v_1{}^{(b)}$	$v_3{}^{(c)}$
Molecular ground state[1]	—	1151	518	1362
$(6a_1)$	12·29(0)	—	400	—
$(1a_2)$ & $(4b_2)$	12·98(0)	—	380	930 ?*
	?	—	500	—
$(5a_1)$ & $(1b_1)$	15·97(2)	850	—	1240 ?*
	16·33(5) or 16·44(6)*	900	—	—

* See text.

(a) symmetric stretching mode   (b) bending mode   (c) antisymmetric stretching mode.

The vibrational structure on the first band (Adiabatic I.P. 12·29(0) eV Vertical I.P. 12·50(2) eV) is poorly resolved (Figure 4.44), and the first two peaks in the series are somewhat asymmetric in shape, the spacing between them (500 cm^{-1}) seeming to be significantly larger than the vibrational spacing for the rest of the series (400 cm^{-1}). It is considered that the series is one in v_2, the bending mode, since strong excitation of this mode would be expected if, as is likely, removal of a $(6a_1)$ electron results in a large change in bond angle (cf. NO_2). Eland and Danby considered that the bond angle increased from 119° to 137° during this ionization process.

The value obtained for the adiabatic I.P. is slightly lower than that recorded by Dibeler and Liston from their photon-impact work,[63] 12·32 eV, but both figures exclude the value of 12·05 eV given by Price[64] from a rought extrapolation of a Rydberg series. There is some evidence for vibrational structure on the photoionization efficiency curve of $SO_2{}^+$,[63] but the resolution is insufficient for the intervals to be measured accurately.

The broad band between 13 and 14 eV (Figure 4.45, Adiabatic I.P. 12·98(0) eV, Vertical I.P.'s 13·2 eV, 13.5 eV) seems to relate to two separate ionization processes. The value for the adiabatic I.P. of the second band is lower than that obtained by Eland and Danby (13·010 eV), and our interpretation of the vibrational structure (Figure 4.45 and Table 4.9) is also different.

The third band, which starts part way through the second, consists of a simple series in v_2 only. We estimate that at least two members of the series are obscured by the second band.

It is considered that these I.P.'s correspond to removal of electrons from the $(4b_2)$ and $(1a_2)$ orbitals, and that some considerable d–π bonding is involved in them, since without it they should be nearly nonbonding orbitals,[59] yet the shape of the photoelectron bands indicates fairly strong bonding character.

These ionic states could not be detected in the photoionization efficiency curve of SO_2^+.[63]

The portion of the spectrum between 16 and 17·5 eV again seems to include two bands (Figure 4.46 Adiabatic I.P.'s 15·97 eV and 06·44(6) or 16·33(5) eV, Vertical I.P. 16·6 eV and 16.6 eV) and not one as might be at first thought. Series in v_1 (see Table 4.9) are excited in both bands, but there is the possibility of an additional series in the fourth band (see Figure 4.46) involving the addition of one quantum of another frequency (possibly v_3) to the members of the first series. The validity of this series has some importance since the peak at 16·33(5) if not a member of it represents the first member of the fifth band series.

Eland and Danby observed a single peak at 16·73 eV, which did not fit into either the fourth or fifth band series, and which they considered possibly represented a further electronic state. We have not observed this peak. Diffuseness in the spectrum above 16·5 eV may be due to rapid fragmentation since it is thought from electron impact work[65,66] that the A.P. of SO^+ lies in this region. Dibeler and Liston's work shows that the A.P. of SO^+ is as low as 15·8 eV, and so diffuseness above 16·5 eV in the photoelectron spectrum cannot be ascribed to the onset of this fragmentation process.

(D) Nitrogen Dioxide and Ozone

The high resolution spectra of nitrogen dioxide and ozone have not yet been obtained, and so the original low-resolution spectra obtained with a retarding-grid instrument[67,68] are included here for the sake of completeness (Figure 4.47), since their electronic structures are closely related to that of sulphur dioxide.

Nitrogen dioxide, a bent triatomic molecule ($\theta = 134°$), has one less valence electron than ozone and sulphur dioxide, and the ground state molecular configuration may be represented as:

$$\ldots (1b_1)^2 \, (5a_1)^2 \qquad (1a_2)^2 \, (4b_2)^2 \qquad (6a_1) \ldots {}^2A_1$$

$\underbrace{\hphantom{(1b_1)^2 \, (5a_1)^2}}$ from $(\pi_u)^4$ of the linear conformation

from $(\pi_g)^4$ of the linear conformation

Removal of the most loosely bound electron, $(6a_1)$, would leave NO_2^+ isoelectronic with CO_2, and so NO^+ in its ground state is expected to be linear with configuration:

$$\ldots (1\pi_u)^4 \, (1\pi_g)^4 \ldots \, ^1\Sigma_g^+$$

Removal of electrons from the deeper orbitals of NO_2 leads to a large number of possible ionic states, owing to the presence of the unpaired electron in the $(6a_1)$ orbital (cf. nitric oxide, Chapter 3).

The value of the first I.P. of nitrogen dioxide is still open to discussion, with previous estimates by electron impact and photon impact measurements ranging from 9·78 to 13·98 eV.[69-74]

The literature values can be roughly separated into two groups, one centred around 10 eV, and the other centred around 11 eV. Some work in fact indicates both values.[71] Collin[75] has favoured the lower value as representing the true first I.P., and 9·78 eV seems to be the most recently accepted low value. This low value could arise from the process.

$$NO_2 \longrightarrow NO + O \quad 9·7\text{--}9·8 \text{ eV}$$

followed by ionization of the nitric oxide produced and (perhaps) capture of O.

$$NO^+ + O \longrightarrow NO_2^+$$

From the grid spectrum of nitrous oxide, the first band appears to start at an I.P. of approximately 10·97 eV, i.e. favouring the higher group of values. The shape of the band is consistent with the removal of an antibonding electron $(6a_1)$ with a large change in geometry ($\theta = 180° \leftarrow 124°$), the vertical I.P. being at least 0·5 eV higher than the adiabatic. There does not seem to be any indication in the spectrum of a lower value for this I.P.

The poor resolution of the spectrum and the obviously large number of ionic states present restrict the deductions that can be made from the rest of the spectrum. The remaining I.P.'s bear little resemblance to values quoted from other determinations, which in any case disagree badly with each other. The one exception is an I.P. of 18·86 eV, marked I_7 which obviously relates to a nearly non-bonding orbital. This is in exact agreement with a Rydberg series limit of 18·87 eV,[76] which was considered to relate to removal of an electron from the $(2a_1)$ orbital. There appears to be a vibrational frequency of ~ 1050 cm^{-1} associated with the band, which may represent the symmetric stretching frequency ν_1 (thought to be 1320 cm^{-1} in the molecular ground state).

A proper analysis of NO_2 and in particular the definite assignment of the first I.P. rests on the obtaining of a good high-resolution spectrum.

Ozone is isoelectronic with sulphur dioxide and can be given the same electronic configuration (see page 84). Detailed MO calculations have been performed on ozone,[77] and the computed orbital energy-levels are given in Table 4.10.

Table 4.10   Calculated orbital energies of O_3[77]

Molecular orbital	Orbital energy (eV)	Orbital character
$(1a_2)$	13·7	Nonbonding
$(6a_1)$	14·8	Weakly antibonding
$(4b_2)$	15·6	Weakly antibonding
$(1b_1)$	21·3	Strongly bonding
$(5a_1)$	22·8	Weakly bonding
$(3b_2)$	25·2	Weakly antibonding
$(4a_1)$	32·6	Strongly antibonding

The grid spectrum shown here[68] is to our knowledge the only one so far obtained. It is thought that the first I.P. lies at 12·52 eV, and probably relates to the removal of a $(6a_1)$ orbital electron, which is weakly bonding or antibonding in a similar fashion to that in sulphur dioxide. Because of the similarity of the structures of sulphur dioxide and ozone it is not considered likely that this I.P. represents the removal of a $(1a_1)$ orbital electron, despite the fact that the MO calculation indicates this to be the highest filled MO. The experimental value is in fair agreement with that of 12·8 eV reported from photon impact work[78] and electron impact work.[79] There is a slight possibility that a lower I.P. exists near 12·3 eV, but it seems probable that any structure here may be due to residual oxygen.

The peak at an I.P. of 13·5 eV in the spectrum probably relates to the $(1a_2)$ and $(4b_2)$ pair of orbitals (cf. sulphur dioxide), and judging from the spectrum they would both have to be fairly nonbonding. This is in agreement with the calculations.

If there is a genuine peak at 12·3 eV (i.e. not due to residual oxygen), then it could relate to the $(1a_2)$ orbital, and the peak at 13·5 eV to the $(3b_2)$ orbital, thus making the experimental order of the I.P.'s the same as the calculated order of orbital energy levels (Table 4.10).

The very broad peak centred at approximately 17 eV probably relates to the removal of electrons from the $(5a_1)$ and $(1b_1)$ orbitals (cf. sulphur dioxide), which are expected to be strongly bonding and thus to give broad, featureless bands.

The remaining peak at 19·3 eV may represent ionization from the $2b_2$ orbital.

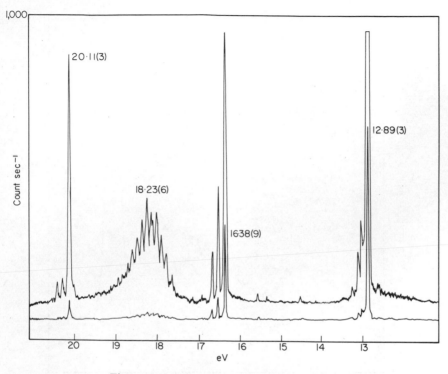

Figure 4.1   Nitrous oxide (full spectrum)

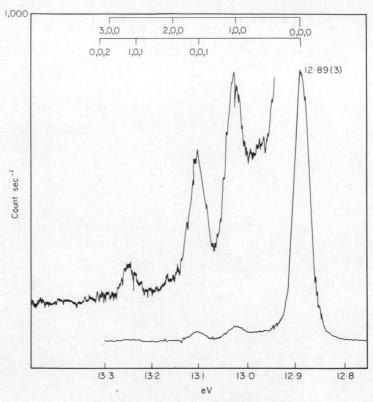

Figure 4.2   Nitrous oxide (first band)

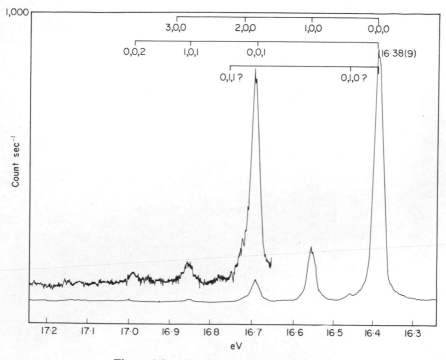

Figure 4.3   Nitrous oxide (second band)

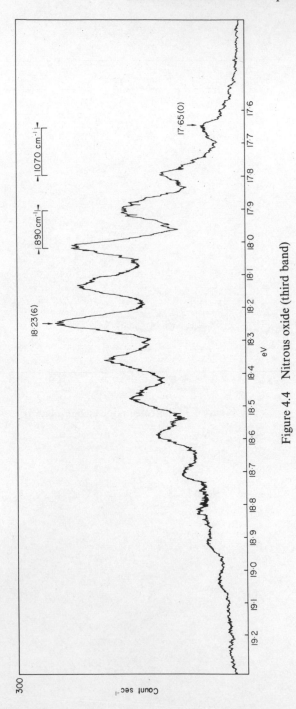

Figure 4.4   Nitrous oxide (third band)

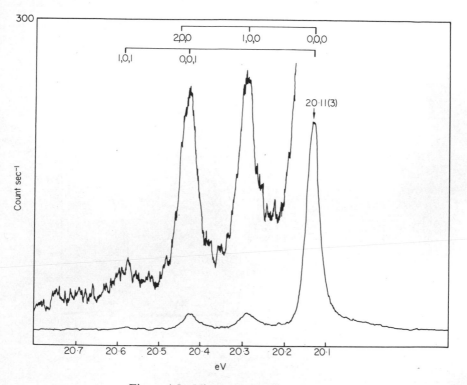

Figure 4.5   Nitrous oxide (fourth band)

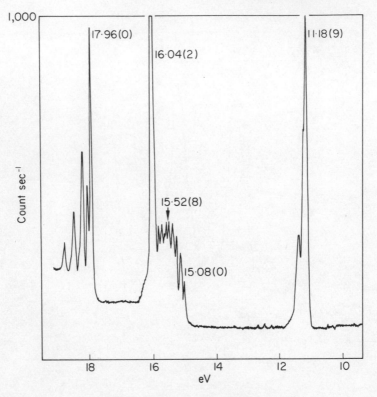

Figure 4.6   Carbonyl sulphide (full spectrum)

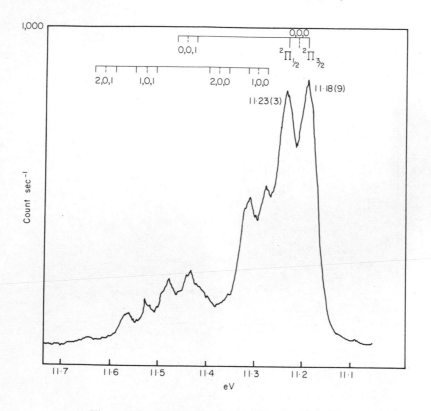

Figure 4.7   Carbonyl sulphide (first band) [R]

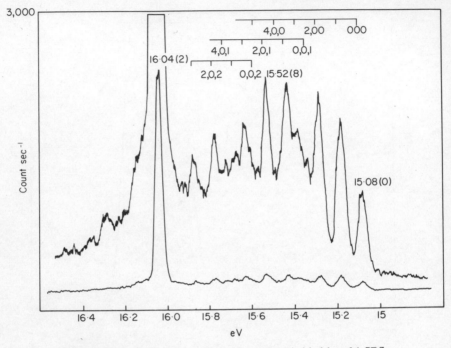

Figure 4.8   Carbonyl sulphide (second and third bands) [R]

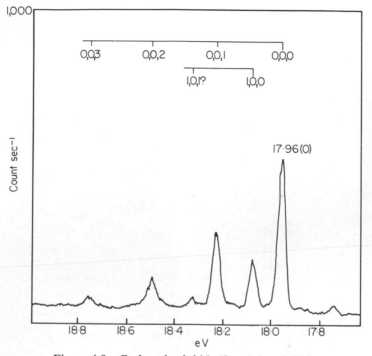

Figure 4.9   Carbonyl sulphide (fourth band) [R]

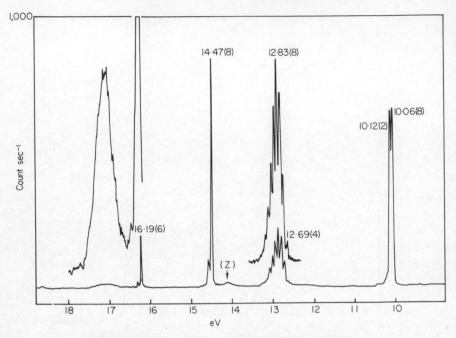

Figure 4.10   Carbon disulphide (full spectrum)

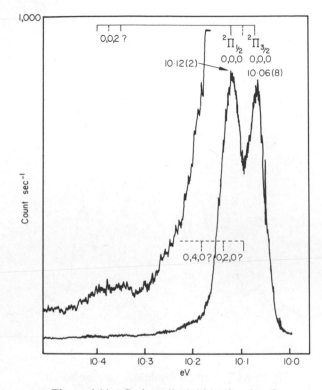

Figure 4.11   Carbon disulphide (first band)

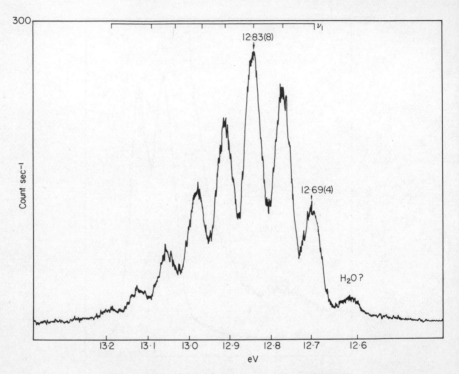

Figure 4.12   Carbon disulphide (second band)

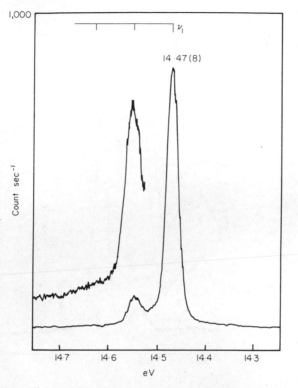

Figure 4.13   Carbon disulphide (third band)

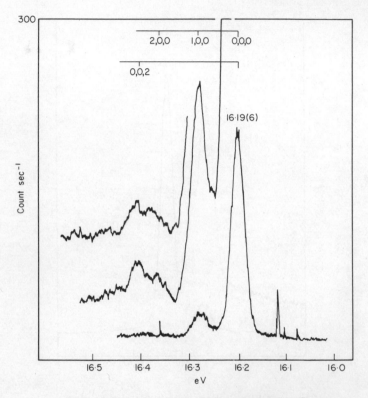

Figure 4.14   Carbon disulphide (fourth band)

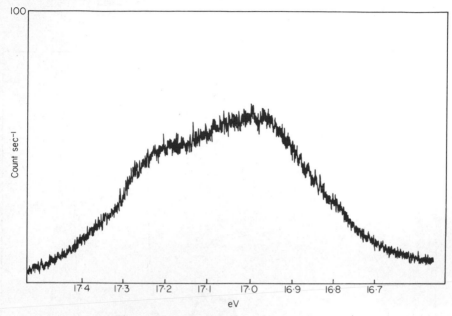

Figure 4.15   Carbon disulphide (fifth band)

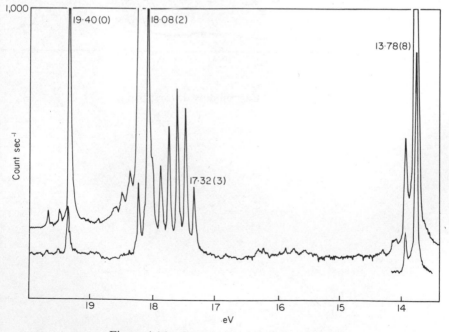

Figure 4.16   Carbon dioxide (full spectrum)

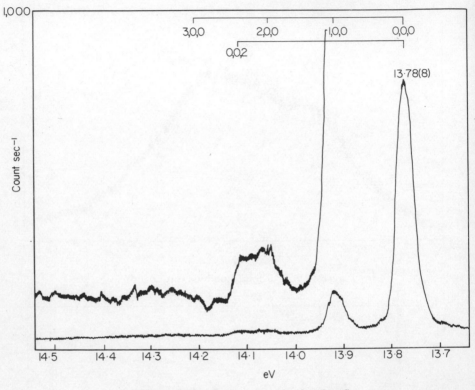

Figure 4.17   Carbon dioxide (first band)

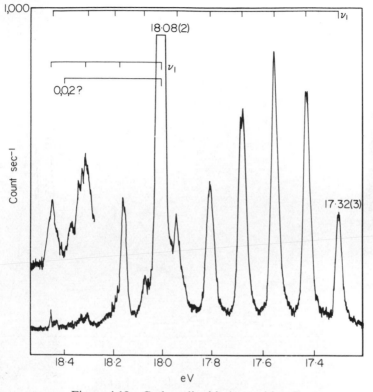

Figure 4.18   Carbon dioxide (second band)

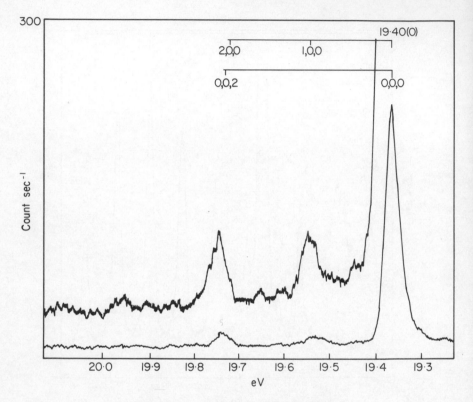

Figure 4.19   Carbon dioxide (third band)

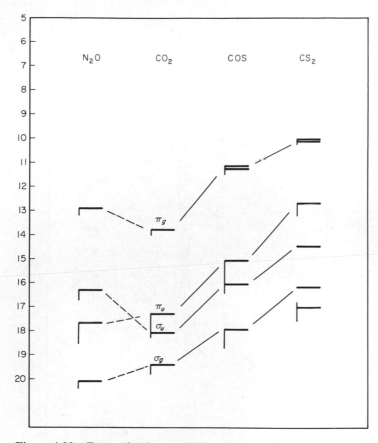

Figure 4.20   Energy-level correlation diagram for the linear triatomic molecules using adiabatic ionization potentials measured
from their photoelectron spectra

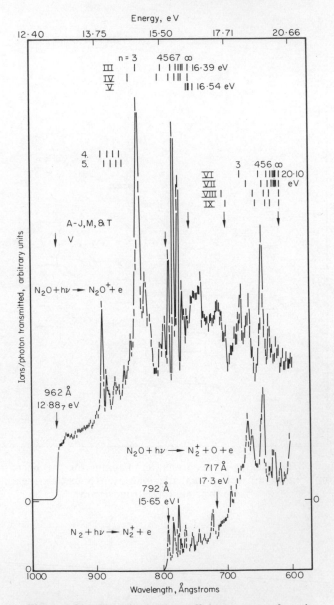

Figure 4.21 Photoionization efficiency curve for nitrous oxide obtained by Dibeler and Walker (reproduced from Reference 9 with permission)

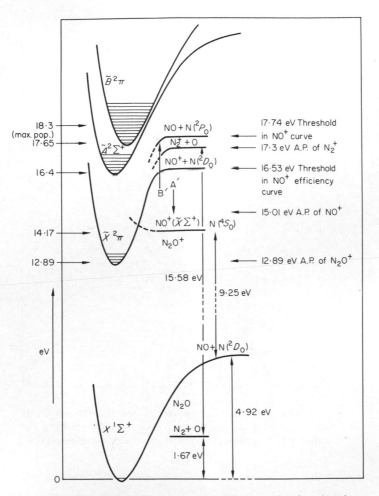

Figure 4.22 Potential energy curves (schematic) for the three lowest electronic states of the nitrous oxide molecular ion deduced from the photoelectron spectrum. The bottom curve is that for the ground-state of the neutral molecule

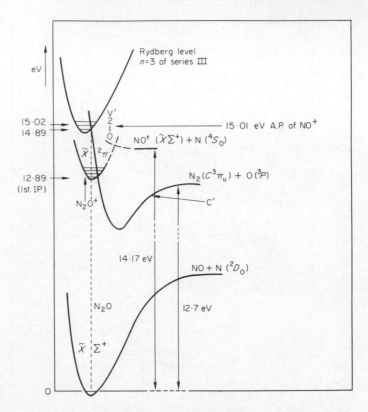

Figure 4.23 Potential energy curves for the nitrous oxide molecular ion in the two lowest energy states showing the crossing of these with an excited molecular state (see text)

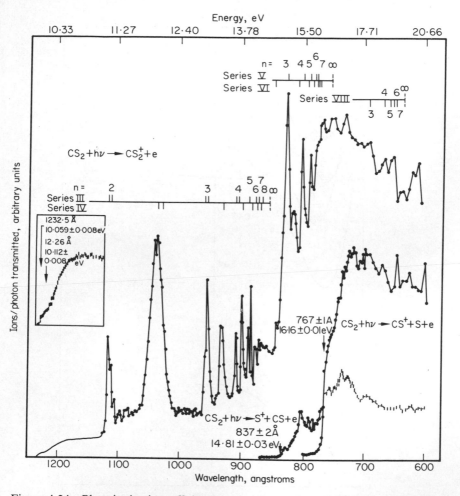

Figure 4.24 Photoionization efficiency curve for carbon disulphide obtained by Dibeler and Walker (reproduced from Reference 9 with permission)

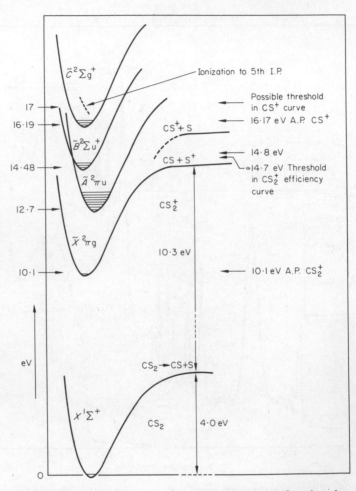

Figure 4.25 Potential energy curves (schematic) for the electronic states of the carbon disulphide molecular ion deduced from the photoelectron spectrum. The bottom curve is that for the ground-state of the neutral molecule

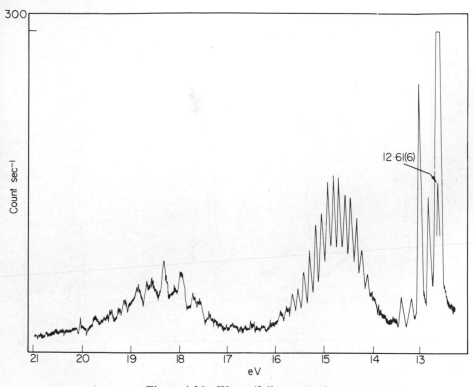

Figure 4.26   Water (full spectrum)

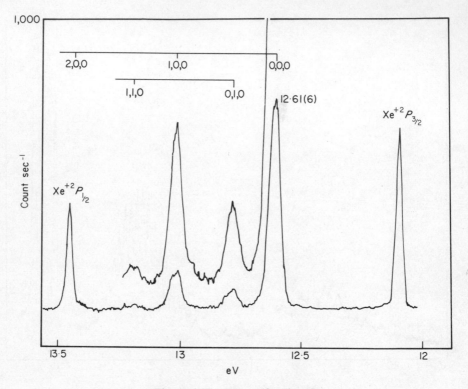

Figure 4.27　Water (first band)

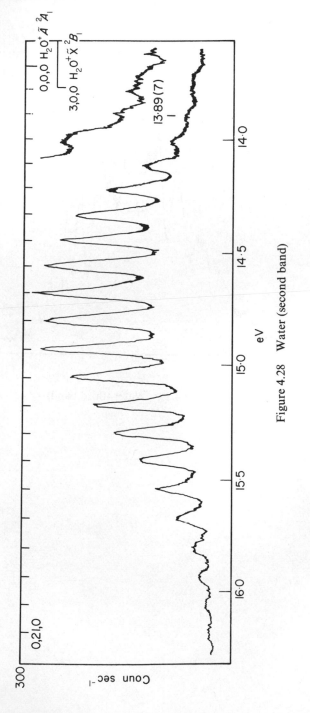

Figure 4.28   Water (second band)

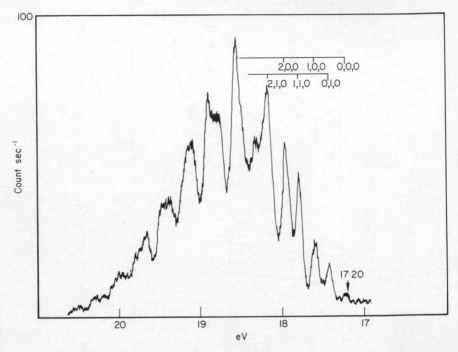

Figure 4.29   Water (third band)

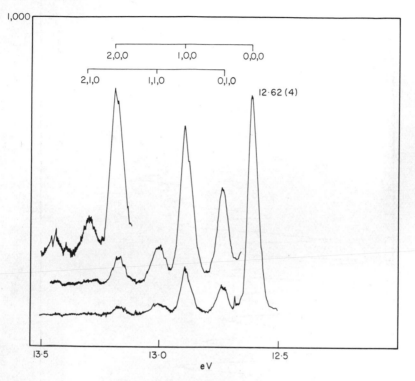

Figure 4.30 Deuterium oxide (first band)

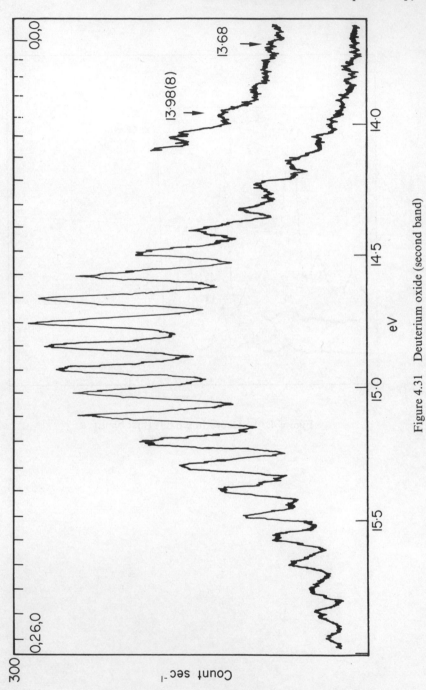

Figure 4.31   Deuterium oxide (second band)

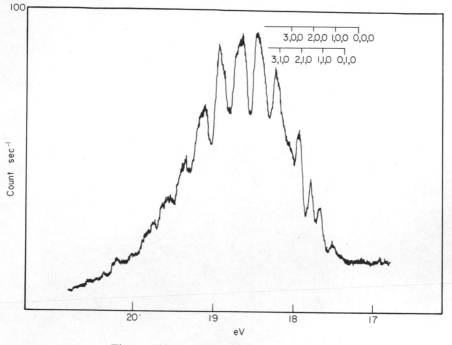

Figure 4.32   Deuterium oxide (third band)

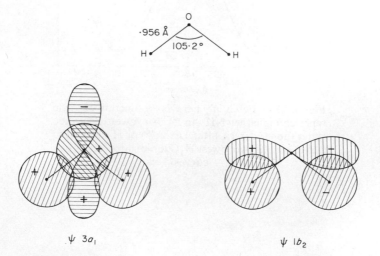

$\psi\ 3a_1$ $\psi\ 1b_2$

Figure 4.33   The occupied orbitals of the water molecules illustrated in terms of their constituent atomic orbitals

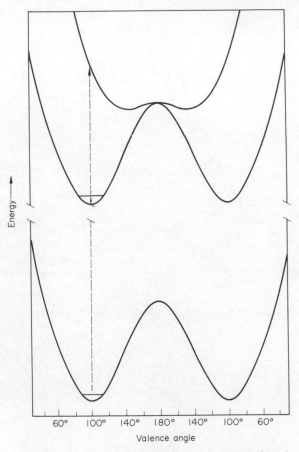

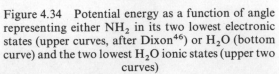

Figure 4.34   Potential energy as a function of angle
representing either NH_2 in its two lowest electronic
states (upper curves, after Dixon[46]) or H_2O (bottom
curve) and the two lowest H_2O ionic states (upper two
curves)

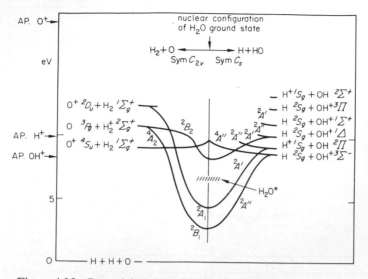

Figure 4.35   Potential energy of H_2O as a function of O—H bond
length (reproduced from Reference 33 with permission)

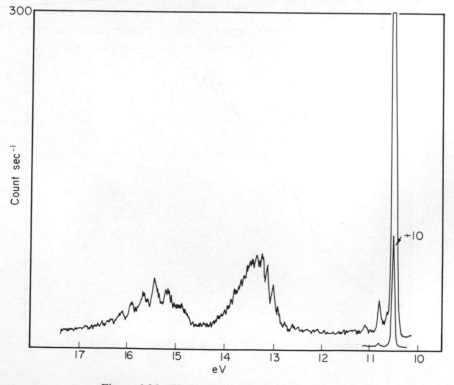

Figure 4.36   Hydrogen sulphide (full spectrum)

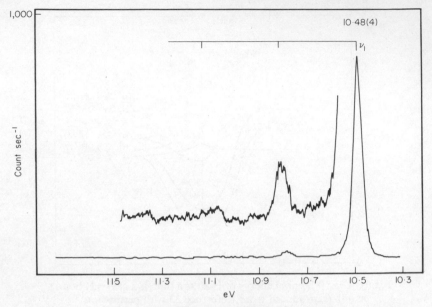

Figure 4.37   Hydrogen sulphide (first band)

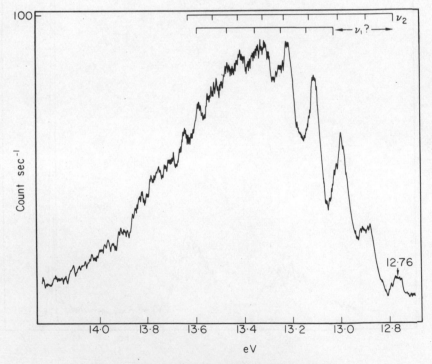

Figure 4.38   Hydrogen sulphide (second band)

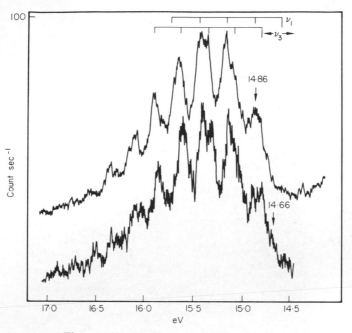

Figure 4.39   Hydrogen sulphide (third band)

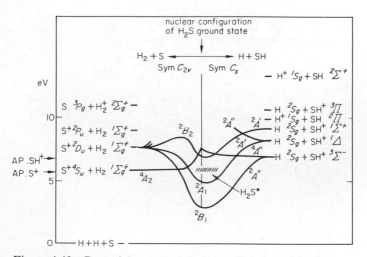

Figure 4.40   Potential energy of H_2S as a function of S—H bond length (reproduced from Reference 33 with permission)

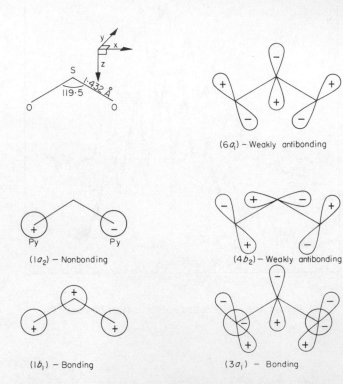

Figure 4.41 The occupied orbitals of the sulphur dioxide molecule illustrated in terms of their constituent atomic orbitals

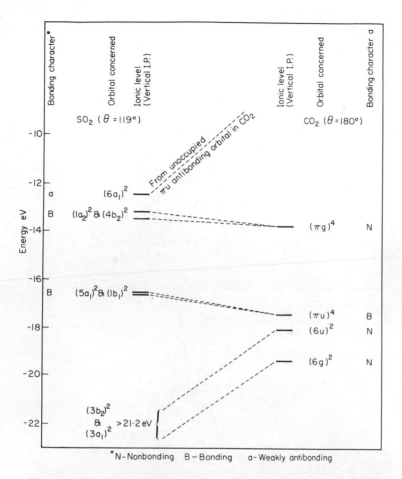

Figure 4.42   The correlation of the energy levels of sulphur dioxide and carbon dioxide from comparison of their photoelectron spectra

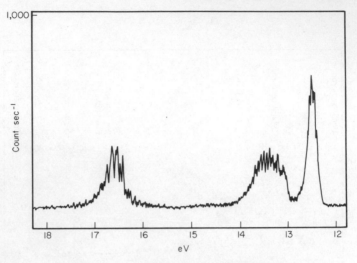

Figure 4.43   Sulphur dioxide (full spectrum)

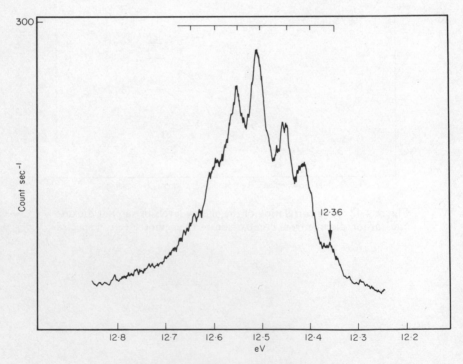

Figure 4.44   Sulphur dioxide (first band)

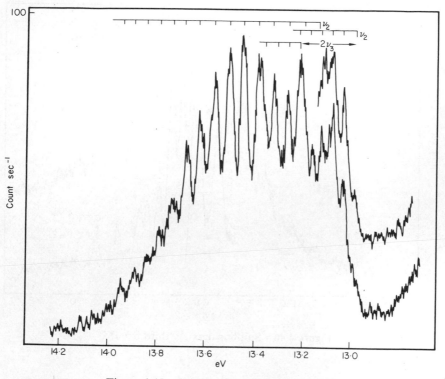

Figure 4.45   Sulphur dioxide (second band)

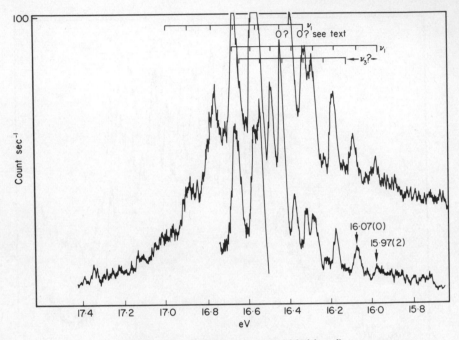

Figure 4.46   Sulphur dioxide (third band)

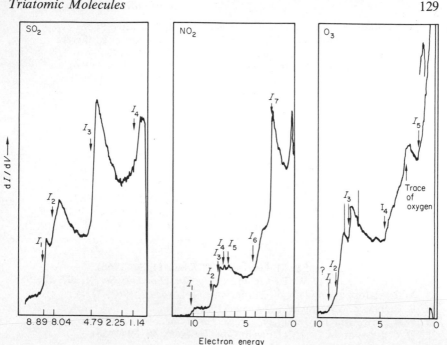

Figure 4.47   Sulphur dioxide, Nitrogen dioxide and Ozone (using a cylindrical-grid energy analyser (reproduced from *Physical Methods in Advanced Inorganic Chemistry*, Ed. H. A. O. Hill and P. Day, Interscience, London, 1968, Chapter 3)

REFERENCES

1. M. I. Al-Joboury, D. P. May and D. W. Turner, *J. Chem. Soc.*, **1965**, 6350.
2. D. W. Turner and D. P. May, *J. Chem. Phys.*, **46**, 1156 (1967).
3. J. H. D. Eland and C. J. Danby, *Intern. J. Mass Spectr. Ion Phys.*, **1**, 111 (1968).
4. C. R. Brundle and D. W. Turner, *Intern. J. Mass Spectr. Ion Phys.*, **2**, 195 (1969).
5. D. Villarejo, R. Stockbauer and M. G. Inghram, *J. Chem. Phys.*, **48**, 3342 (1968).
6. J. E. Collin and P. Natalis, *Intern. J. Mass Spectr. Ion Phys.*, **1**, 121 (1968).
7. P. Natalis and J. E. Collin, *Chem. Phys. Letters*, **2**, 79 (1968).
8. Y. Tanaka, A. S. Jursa and F. J. Leblanc, *J. Chem. Phys.*, **32**, 1199, 1205 (1960).
9. V. H. Dibeler and J. A. Walker, International Mass Spectrometry Conference, Berlin, September 1967.
10. J. H. Callomon, *Proc. Chem. Soc.*, **1959**, 33.
11. J. D. Carette, *Can. J. Phys.*, **45** 2931 (1967).
12. G. Herzberg, *Molecular Spectra and Molecular Structure*, Part III D, Van Nostrand, New York, 1967.
13. M. Haugh, T. G. Slanger and K. D. Bates, *J. Chem. Phys.*, **44**, 838 (1966).
14. F. M. Matsunga and K. Watanabe, *J. Chem. Phys.*, **46**, 4457 (1967).
15. S. Leach, *J. Chim. Phys.*, **61**, 1493 (1964).
16. V. Čermák, *J. Chem. Phys.*, **44**, 3781 (1966).

17. J. H. Callomon, *Proc. Roy. Soc.* (*London*), *Ser. A*, **244**, 220 (1968).
18. S. Mrozowski, *Phys. Rev.*, **72**, 691 (1947), and references cited therein.
19. V. A. Koryoshkin, *Dokl. Akad. Nauk. SSSR*, **167**, 1035 (1966).
20. T. E. Sharp and H. M. Rosenstock, *J. Chem. Phys.*, **41**, 3453 (1964).
21. J. Collin, *Mem. Soc. Roy. Sci. Liege, Vol XIV, Hors Ser.* **5**, 1 (1967).
22. W. C. Johns, *Can. J. Phys.*, **42**, 1004 (1964).
23. Y. Tanaka and M. Ogawa, *Can. J. Phys.*, **40**, 879 (1962)
24. R. S. Mulliken, *J. Chem. Phys.*, **3**, 720 (1935).
25. J. F. Mulligan, *J. Chem. Phys.*, 347 (1951).
26. A. D. McLean, *J. Chem. Phys.*, **32**, 1595 (1960).
27. S. D. Peyerimhoff, R. J. Buenker and J. L. Whitten, *J. Chem. Phys.*, **46**, 1707 (1967).
28. E. Clementi, *J. Chem. Phys.*, **36**, 750 (1962).
29. R. Spohr and E. Von Puttkamer; *Z. Naturforsch.*, **22**A, 705 (1967).
30. G. R. Cook, P. H. Metzger and M. Ogawa, *J. Chem. Phys.*, **44**, 2935 (1966).
31. G. R. Cook, P. H. Metzger and M. Ogawa, *J. Opt. Soc. Am.*, **58**, 129 (1968).
32. H. J. Lempka, T. R. Passmore and W. C. Price, *Proc. Roy. Soc.* (*London*), *Ser. A*, **304**, 53 (1968).
33. F. Fiquet-Fayard and P. M. Guyon, *Mol. Phys.*, **11**, 17 (1966).
34. C. R. Brundle and D. W. Turner, *Proc. Roy. Soc.* (*London*), *Ser. A*, **307**, 27 (1968).
35. M. I. Al-Joboury and D. W. Turner, *J. Chem. Soc. B*, **1967**, 373.
36. B. Brehm, *Z. Naturforsch.*, **21**A, 196 (1966).
37. W. C. Price and T. M. Sugden, *Trans. Farad Soc.*, **44**, 108 (1948).
38. D. C. Frost and C. A. McDowell, *Can. J. Chem.*, **36**, 39 (1958).
39. M. Cottin, *J. Chem. Phys.*, **56**, 1024 (1959).
40. W. C. Price, *J. Chem. Phys.*, **4**, 147 (1959).
41. R. Botter and H. M. Rosenstock, *International Conference of Mass Spectroscopy*, Berlin, 1967. (Advances in Mass Spectrometry, Vol. 4.)
42. W. C. Johns, *Can. J. Phys.*, **41**, 209 (1963).
43. M. Krauss, *J. Res. Nat. Bur. Std.*, *A*, **68**, 635 (1964).
44. R. Botter and H. M. Rosenstock, Private communication to be published.
45. K. Dressler and D. A. Ramsey, *Phil. Trans. Roy. Soc.* (*London*), *Ser. A*, **241**, 553 (1959).
46. R. N. Dixon, *Mol. Phys.*, **9**, 359 (1965).
47. V. H. Dibeler, J. A. Walker and H. M. Rosenstock, *J. Res. Nat. Bur. Std.*, *A*, **70**, 459 (1966).
48. A. D. Baker, C. R. Brundle and D. W. Turner; *Intern. J. Mass Spec. Ion Physics*, **1**, 443 (1968).
49. M. I. Al-Joboury and D. W. Turner, *J. Chem. Soc.*, 4434 (1964).
50. R. Moccia, *J. Chem. Phys.*, **40**, 2186 (1964).
51. D. G. Carroll, A. T. Armstrong and S. P. McGlynn, *J. Chem. Phys.*, **44**, 1865 (1966).
52. S. D. Thompson, D. G. Carroll, F. Watson, M. O'Donnell and S. D. McGlynn, *J. Chem. Phys.*, **45**, 1367 (1966).
53. M. I. Al-Joboury, *Ph.D. Thesis*, London University, 1964.
54. H. M. Rosenstock, *Advan. Mass Spectrometry*, *Proc. Conf. Univ. London*, **4**, 523 (1968).
55. W. C. Price, *J. Chem. Phys.*, **4**, 147 (1936).
56. V. H. Dibeler and S. K. Liston, *J. Chem. Phys.*, **49**, 482 (1968).
57. D. C. Frost and C. A. McDowell, *Can. J. Chem.*, **36**, 39 (1958).

58. V. H. Dibeler and H. M. Rosenstock, *J. Chem. Phys.*, **39**, 3106 (1963).
59. R. S. Mulliken, *Can. J. Chem.* **36**, 10 (1958).
60. S. P. Ionov, M. A. Porai-Koshits and G. V. Tsintsadze, *Soobshch. Akad. Nauk Gruz. SSR*, **35**, 559 (1964).
61. S. P. Ionov and M. A. Porai-Koshits, *Zh. Strukt. Khim.*, **7**, 252 (1966).
62. J. H. D. Eland and C. J. Danby, *Intern. J. Mass Spectr. Ion Phys.*, **1**, 111 (1968).
63. V. H. Dibeler and S. K. Liston, *J. Chem. Phys.*, **49**, 482 (1968).
64. W. C. Price, *J. Chem. Phys.*, **4**, 147 (1936).
65. F. H. Field and J. L. Franklin, *Electron Impact Phenomena*, Academic Press, New York, 1957.
66. R. M. Reese, V. H. Dibeler and J. L. Franklin, *J. Chem. Phys.*, **29**, 880 (1958).
67. M. I. Al-Joboury and D. W. Turner, *J. Chem. Soc.*, **1964**, 4434.
68. T. W. Radwan and D. W. Turner, *J. Chem. Soc., A*, **1966**, 85.
69. J. E. Collin and F. P. Lossing, *J. Chem. Phys.*, **28**, 960 (1958).
70. G. L. Weissler, J. A. R. Samson, H. Ogawa and G. R. Cook, *J. Opt. Soc. Am.*, **49**, 338 (1959).
71. T. Nakayama, M. Y. Kitamura and K. Watanabe, *J. Chem. Phys.*, **30**, 1180 (1959).
72. R. W. Kiser, *Comp. Tables of Ion Potls*, United States Atomic Energy Commission TID 6142, 1960, p. 147.
73. R. W. Kiser and I. C. Hisatsure, *J. Phys. Chem.*, **65**, 1444 (1961).
74. R. J. Kandel, *J. Chem. Phys.*, **23**, 84 (1955).
75. J. E. Collin, *Bull. Soc. Roy. Sci. Liege*, **1963**, 133.
76. Y. Tanaka and A. S. Jursa, *J. Chem. Phys.*, **36**, 2493 (1962).
77. I. Fischer-Hjalmars, *Arkiv Fysik*, **11**, 529 (1957).
78. I. Omura, *Bull. Res. Inst. Appl. Elec.*, **6**, 15 (1954).
79. J. T. Herron and H. I. Shiff, *J. Chem. Phys.*, **24**, 1266 (1956).

CHAPTER 5

Formaldehyde and Related Compounds Formamide and Formic Acid

1. FORMALDEHYDE AND ITS DEUTERIUM ANALOGUES

(A) Formaldehyde

The four bands in the spectrum of formaldehyde[1] (Figures 5.1–5.4) indicate four orbitals of energy greater than $-21 \cdot 2$ eV. The values for the vibrational frequencies in each state of the ion are collected in Table 5.1, together with the I.P. values and the ground state molecular frequencies.[2]

Table 5.1 Ionization potentials and vibrational frequencies of formaldehyde

Ionic state	Photoelectron band	Adiabatic I.P. (eV)	Vibrational frequencies (cm^{-1})		
			$v_1{}^{(a)}$	$v_2{}^{(b)}$	$v_3{}^{(c)}$
$^2B_2$	1st	10·88(4)	2560	1590	1210
$^2B_1$	2nd	14·09(5)	(1400?)	1210	—
$^2B_2$	3rd	15·85(4)	—	1270	1270?
$^2A_1$	4th	16·25(4)	1400?	—	—
Ground molecular state—$^1A_1{}^2$	—	—	2780	1744	1503

(a) C—H stretching mode.   (b) C=O stretching mode.   (c) $\begin{array}{c}H \\ \diagdown \\ \diagup \\ H\end{array}$ C deformation mode.

From the results it is clear that breaks reported in electron impact efficiency curves[3] at 11·8 eV and 13·1 eV cannot be ascribed to direct ionization. There is now ample evidence [4, 5] to suggest that both these breaks are ascribable to autoionization processes.

Formaldehyde-d$_2$ gives a spectrum (Figures 5.5–5.7) whose overall form is similar to that of H$_2$CO, but with differences in the vibrational structure, particularly in the third band (Figure 5.7). The I.P.'s and vibrational frequencies are recorded in Table 5.2.

Table 5.2 Ionization potentials and vibrational frequencies of d_2-formaldehyde

Ionic state	Photoelectron band	Adiabatic I.P. (eV)	Vibrational frequencies		
			$v_1^{(a)}$	$v_2^{(b)}$	$v_3^{(c)}$
$^2B_2$	1st	10·90(4)	1910	1560	870
$^2B_1$	2nd	14·09(5)	(1400?)	1210	—
$^2B_2$	3rd	15·84(6)	—	1270	940?
$^2A_1$	4th	?	990?	—	—
Ground molecular state—$^1A_1$[14]	—	—	2056	1700	1106

(a) C—D stretching mode.   (b) C=O stretching mode.   (c)

$$\begin{array}{c} D \\ \diagdown \\ C \\ \diagup \\ D \end{array} \text{ deformation mode.}$$

Calculated eigenvalues for the molecular orbitals involved, which are illustrated schematically in Figure 5.8, have varied widely as the methods of computation improved. A summary of recent calculations is given in Table 5.3.

The ionization potentials indicated by the first band (Figures 5.2, 5.5—Adiabatic I.P. $H_2CO = 10·88(4)$ eV $D_2CO = 10·90(4)$ eV) are in agreement with previous determinations.[6,7] All three totally symmetric vibrational modes are excited in the ground state of the ion (Tables 5.1, 5.2), and the similarity of the frequencies compared to those of the molecule in its ground

Table 5.3 Calculated eigenvalues of the molecular orbitals of formaldehyde (eV)

	ψ_1	ψ_2	ψ_3	ψ_4	ψ_5	ψ_6
Goodfriend and coworkers,[9] 1960	11·53	15·08	19·25	22·10	21·94	38·73
Foster and Boys,[10] 1960	10·50	12·80	15·55	18·38	22·80	37·35
Peters,[11] 1963	10·4	12·8	15·5	18·5	23·0	37·0
Carrol and coworkers[12]	10·9	15·3	14·4	16·7	20·9	30·9
	11·0	13·2	14·5	16·6	21·0	30·9
Newton and Palke[13]	10·76	13·54	16·16	18·40	22·62	38·10

state, together with the strong relative intensity of the 0,0,0, component band, demonstrates clearly that an essentially nonbonding electron $(2b_2)$ has been removed with little resultant change in molecular dimensions.

The second bands of formaldehyde and deuteroformaldehyde (Figures 5.3, 5.6 Adiabatic I.P. H_2CO, D_2CO 14·09(5) eV, Vertical I.P. H_2CO, D_2CO 14·38(8) eV) consist of a series of doublets, the lengths of the series indicating that a strongly bonding electron has been removed $(1b_1$, CO, π bonding). The main series converges slightly to higher ionization energies. The interpretation of this band has caused some difficulty,[1, 8] but it would seem that the vibrational frequencies of both the modes excited in D_2CO^+ are identical to those in H_2CO^+, only the relative intensities having changed.

The third band in the spectrum of formaldehyde (Figure 5.4, Adiabatic I.P. H_2CO 15·85(4) eV) consists of a single slightly converging series of narrow peaks. An examination of the third band of formaldehyde-d_2 (Figure 5.7, Adiabatic I.P. D_2CO 15·84(6) eV) shows this simplicity to be the result of an equality of frequencies for two vibrational modes. The normal modes of vibration for X_2CO are indicated in Figure 5.9. On deuteration the frequency of one mode, v_2, remains unchanged, while the other is reduced. The latter frequency obviously relates to a C—H mode, since the reduction on deuteration (1·36:1) is exactly that expected. It is considered likely that this mode is v_3, the deformation mode, rather than v_1 the stretching mode, since the reduction required for it to be v_1 is rather large (see Tables 5.1, 5.2). Since both vibrational modes are excited equally strongly in this state of the ion, and both are reduced in frequency compared with the molecular ground-state frequencies, it seems most probable that the ion results from the removal of an electron from a bonding orbital covering the whole molecule, namely ψ_4 $(1b_2)$.

The fourth band of formaldehyde (Figure 5.4 Adiabatic I.P. 16·25) starts under the third, and it is possible that the adiabatic I.P. is lower than the value quoted. The band becomes complex to higher ionization energies, and in formaldehyde-d_2 (Figure 5.7), the only features which can be picked out are three peaks at 16·74 eV, 16·86 eV, and 16·99 eV, having a separation corresponding to a vibrational frequency of approximately 990 cm^{-1}. This reduction in frequency on deuteration (see Tables 5.2, 5.3) is about the expected value for a C—H vibrational mode, and it is found that if the series is extrapolated to lower ionization energies, a peak would be positioned almost coincident with the adiabatic I.P. in formaldehyde at 16·24 eV.

It is likely that this state of the ion results from the removal of a $3a_1$ (ψ_3) strongly CH_2 bonding electron, mode v_1 being excited with the expected large reduction in frequency.

The complexity of the band towards higher ionization energies may be due to peak broadening arising from a short lifetime of the ion. Potential energy curve crossing to another state may be involved though it is relevant that Guyon[5] considers this state of the ion to have a completely repulsive potential surface, with respect to vibrations maintaining the C_{2v} symmetry of the molecule.

A fifth band near 20 eV was detected in studies by May[15] and Turner using a cylindrical grid retarding field and again using a 180° magnetic deflexion analyser. It is not revealed in the spectra recorded here and this may be due in part to the use of the 'analyser scan' mode in which the apparatus is relatively insensitive for the lowest energy electrons. The absence of sharp peaks is probably significant, however, indicating that there is no long-lived state of the ion near 20 eV. This state may well be produced by the loss of a strongly bonding electron and the corresponding Franck–Condon envelope seems to extend over several electron volts, contributing to the difficulty in detecting this band.

(B) Monodeuteroformaldehyde (HDCO)

This spectrum whilst generally similar to that for D_2CO shows differences in the vibrational fine structure (Figures 5.10–5.13) ascribable to its lower symmetry. Whereas in H_2CO and D_2CO only the totally symmetric modes v_1, v_2 and v_3 may be excited in HDCO v_5 and v_6 are allowed as well.

The I.P.'s and vibrational frequencies are summarized in Table 5.4.

Table 5.4   Ionization potentials and vibrational frequencies of formaldehyde-d_1

Ionic state	Photoelectron band	Adiabatic I.P. (eV)	Vibrational frequencies		
			v_1	v_2	v_3
$^2B_2$	1st	10·89(5)	1940	1610	1120
$^2B_1$	2nd	14·09(2)	(1400?)	1210	—
$^2B_2$	3rd	15·83(6)	—	1270*	940?
$^2A_1$	4th	Not known	1060	—	—
Ground molecular state—$^1A_1$[14]	—	—	2121	1723	1400

* v_6 probably also excited with same frequency.

2. FORMAMIDES (Formamide, *N*-Methylformamide, *N,N*-Dimethylformamide)

Special importance is attached to the understanding of the electronic structure and electronic spectrum of the amide group because polypeptides may be regarded as weakly π electron coupled amide residues, their optical properties being closely related to those of the monomeric amide group.

With this in mind M. B. Robin and coworkers [16,17] studied in detail the optical spectra of several carboxylic acids, amides and acyl fluorides. They also carried out a series of MO calculations on these compounds. The calculations showed that in the amides the two highest occupied orbitals, n and π (n refers to the nonbonding σ orbital largely centred on the oxygen atom, i.e. the oxygen lone-pair. The π refers to an orbital largely centred on the N and O atoms) were fairly close lying, and raised the question of to which orbital the first I.P. related. The calculations indicated that for formic acid and formyl fluoride the highest filled orbital is n and hence by Koopmans' Theorem, neglecting reorientation energy, the first I.P. relates to the removal of an n electron. However for formamide the π orbital is placed above the n orbital and Koopmans' Theorem thus indicates that the first I.P. involves the removal of a π electron. When reorientation energy is allowed for, by calculating the I.P.'s by subtracting the total SCF energy of the ion from that of the neutral molecule, it was found that the first I.P. of all three compounds involved the n orbital.

Thus there is little doubt, from the calculations, that for acids and acyl fluorides the first I.P. involves ionization from the n orbital, but the origin of the first I.P. of formamide remains undecided. The actual values found for formamide by the two methods were:

Koopmans' Theorem	1st I.P.	11·32 eV—π	
	2nd I.P.	11·86 eV—n	
SCF energy of ion-SCF	1st I.P.	8·80 eV—n	
energy of neutral molecule	2nd I.P.	9·74 eV—π	

The photoelectron spectra of *N*-methylformamide, and *N,N*-dimethylformamide help to establish to which orbitals the first and second I.P.'s of formamide relate.

The spectra are shown in Figures 5.14–5.21 and the observed I.P.'s and vibrational frequencies are indicated also. The two bands corresponding to the first and second I.P.'s obviously overlap, and appear to change their relative positions in the spectrum on going from formamide through *N*-methylformamide to *N,N*-dimethylformamide.

The vibrational interval of approximately 1600 cm^{-1} associated with one of the bands, and which seems to appear in all three compounds, is very probably the O—C—N antisymmetric stretching mode which in the neutral molecule has a frequency of 1680 cm^{-1},[18] and is describable largely in terms of the stretching of the C—O bond.

The second band in formamide (Figure 5.14) has a vibrational spacing of approximately 600 cm^{-1} associated with it. The mode excited could be the O—C—N deformation (600 cm^{-1} in the molecule[18,19]), or one of the several modes involving the NH$_2$ group which have been assigned to different frequencies by different authors.[19–21] Whichever mode is involved, the band shape changes markedly on going to the methyl compounds.

The simplest explanation is that the band with the simple C—O vibration associated with it relates to the removal of an electron from the *n* orbital and the one changing with the introduction of the methyl groups relates to the *π* orbital (cf. HCN, CH$_3$CN, CH$_3$NC, Chapter 13). If this explanation is correct, it can be deduced that the first I.P. of formamide does relate to the *n* orbital.

The experimental values of the first and second I.P.'s of formamide are, from the photoelectron spectrum:

$$\text{1st I.P.} \quad (n) = 10 \cdot 13 \text{ eV}$$
$$\text{2nd I.P.} \quad (\pi) \leqslant 10 \cdot 51 \text{ eV}$$

The positions of the start of this band is not unambiguously established.

That these are both higher than the calculated values (reorientation energy allowed), is readily explainable as being due to the larger correlation error in the calculated ground-state energy, due to its having one more electron pair than the ion. The calculated values seem to have overallowed for the difference in reorientation energy between the first I.P. and the second I.P., in that the calculated I.P.'s are 0·96 eV apart, whereas the experimental difference is 0·38 eV.

Assuming this interpretation of the origin of the first two photoelectron bands to be correct, the introduction of methyl groups onto the nitrogen atom is seen to be exerting the expected hyperconjugative effect of pushing the *π* orbital to higher energy, with a much smaller effect on the *n* orbital, thus reversing the origin of the first and second I.P.'s.

The third band in the photoelectron spectra of the compounds can reasonably be ascribed to the removal of an electron from the second highest filled *π* level, the introduction of methyl groups in the *N*-methyl and *N,N*-dimethyl compounds again exerting the expected hyperconjugative effect (Figure 5.22).

Nothing definite can be deduced from the other bands in the spectrum, the vibrational spacings could correspond to any number of modes, but it does seem that peaks due to ionization from the methyl orbital are showing up in the 14–15 eV region in the *N*-methyl and *N,N*-dimethyl spectra. This is in the energy region where ionization from methyl-group orbitals often occurs.

The photoelectron spectroscopy results have prompted Robin[22] to re-examine the optical spectra of these compounds, and he has now been able to find nearly identical Franck–Condon profiles in some of the Rydberg bands.

3. Formic Acid

Formic Acid is isoelectronic with formamide and would be expected to have a somewhat similar photoelectron spectrum (Figures 5.23–5.27). The calculations of Robin and coworkers[17] indicated that the highest filled orbital was that of the oxygen lone-pair electrons ($n\sigma$ orbital) and that the ground ionic state was that in which an $n\sigma$ electron had been removed. Thus there is no ambiguity as in the case of formamide (see page 136). The orbital energies obtained were:

Orbital	$10a'$	$2a''$	$9a'$	$1a''$	$8a'$	$7a'$	$6a'$
Type	$n\sigma$	π	π				
Energy (eV)	$-13\cdot01$	$-13\cdot6$	$-16\cdot4$	$-17\cdot8$	$-19\cdot5$	$-20\cdot2$	$-25\cdot0$
Experimental Vertical I.P. (eV)	$11\cdot51$	$12\cdot5$	$14\cdot7$	$15\cdot7$	$17\cdot1$	$17\cdot4$?	

The comparison with the experimental vertical I.P.s is made arbitrarily as the ordering of ionic levels may be different from that of the calculated orbitals, and also it is not completely certain just how many separate bands may be present in the spectrum. Also the possibility cannot be excluded that some of the observed spectral features may be due to the presence of formic acid dimer. The ionization potentials and the observed vibrational frequencies are recorded in Table 5.5.

Table 5.5 Ionization potentials and vibrational frequencies of the ionic states of HCOOH

Photoelectron Band	Adiabatic I.P. (eV)	Vibrational frequency (cm^{-1})					
		v_1	v_2	v_3	v_4	v_5	v_6
Ground molecular state[26]	—	3570	2943	1770	1387	1229	1105
1st	11·33(2)			1450	1300?		
2nd	12·36(7)			2020?		1120	900?
3rd & 4th	≃14·20					950?	950?
5th	16·96(9)			1290?		1290?	

The adiabatic first I.P. (11·33(2) eV, Vertical I.P. 11·51 eV) may be compared to previous estimates of 11·33 eV[23] and 11·51 eV[24] from absorption spectroscopy and electron impact work. The band (Figure 5·24) is very similar to the first band of formamide in shape and vibrational structure. There does seem to be some evidence for the excitation of a second vibrational mode as well as the C=O stretching mode (v_3). This could be the C—H (v_4) or O—H (v_5) in plane bending vibration. The similarity of the general appearance of the band to that in formamide, together with the fact that in formic acid there is little doubt that the first I.P. represents removal of an n electron, would seem to support our speculation that the first band in formamide also relates to the n orbital.

The second band (Figure 5.25) is more complex than the first, as is the case in the amides, and must relate to the removal of an electron from the next highest populated orbital, $2a''$ (π) (Adiabatic I.P. 12·36(7) eV, Vertical I.P. 12·5 eV). At least two vibrational modes are excited. It is possible that these are v_5 (OH bend) and v_6 (C—O stretch). An alternative assignment could be v_5 and v_3. The broadening of the components to high energies could be due to dissociation, for it is reported that the A.P. of CHO^+ lies in the 13 eV region.[25]

Two poorly resolved series may be observed in the third and fourth bands (Figure 5.26) with the likelihood that the mode or modes involved are v_5 and v_6.

The peculiar structure of the fifth band (Figure 5.27) (Adiabatic I.P. 16·96(9) eV Vertical I.P. 17·1 eV) suggests either dissociative broadening occurring around 17·4 eV, a very confused vibrational structure, or the presence of more than one band. The obvious vibrational frequency is probably v_5, or v_3 greatly reduced, and from the two other rather doubtful series picked out two additional frequencies of 1850 cm^{-1} and 2300 cm^{-1} are obtained.

4. KETENE

The photoelectron spectrum of ketene (Figure 5.28) shows at least four bands, all exhibiting some degree of fine structure. The first ionization potential only has been observed,[27] from the convergence of a Rydberg series, at 9·60(5) $\pm$ 0·02 eV.

The first band in the spectrum, at (adiabatic) I.P. 9·64 eV, probably relates to electron loss from the oxygen nonbonding orbital, although Price[27] assigns the first ionization as being due to electron removal from a π orbital localized on the carbon atoms. Similarly, recent MO calculations[28] describe the highest occupied π orbital ($2b_1$) as being CCO bonding. The fine structure associated with the first spectral band (Figure 5.29) can be interpreted in terms of a short series of peaks in v_2, the C=O stretching mode, with a frequency of 2140 cm^{-1}, this being very close to the molecular frequency of 2152 cm^{-1}. The excitation of a second mode, probably the C=C stretching mode v_4, occurs with a frequency of 1020 cm^{-1}.

The third band (Figures 5.30, 5.31) is very similar to the second band in the formaldehyde spectrum (Figure 5.30), and occurs at approximately the same energy. The fine structure consists of a long series with peak separations corresponding to a vibrational frequency of 940 cm^{-1}. The assignment of this is not possible, *a priori*, without an investigation of ketene-d_2, but a comparison with the molecular orbitals of formaldehyde suggests that the ketene orbital is CCO π bonding, and hence we expect the vibrational mode excited to be v_4. On the basis of this length of the progression it is possible that v_3 is the correct assignment, since the large reduction in frequency is then acceptable.

The fourth band, at adiabatic I.P. 16·08 eV (Figure 5.31) contains a vibrational progression corresponding to a frequency of 1020 cm^{-1}. The strongly bonding nature of this orbital indicated by the length of the progression precludes the mode excited from being v_4 since this has a molecular frequency of only 1120 cm^{-1}. This also is more likely to be v_3, the CH$_2$ deformation, reduced from its frequency of 1386 cm^{-1} in the molecule.

The fine structure appears to stop at about 16·7 eV. This may be explained by the 16–18 eV region of the spectrum actually consisting of two overlapping bands, or by the dissociation of the CH$_2$CO$^+$ ion at this energy.

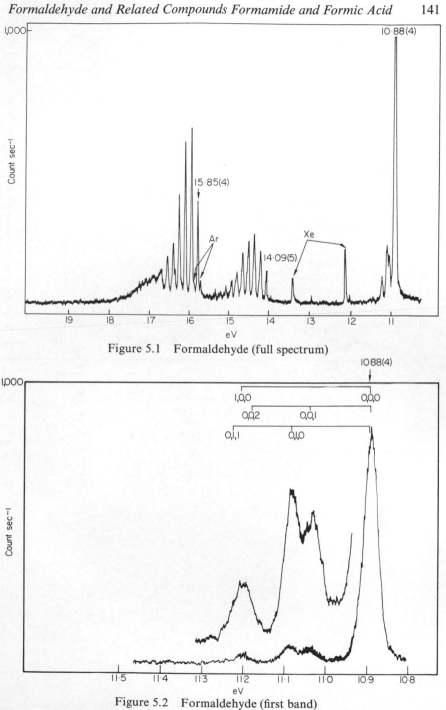

Figure 5.1   Formaldehyde (full spectrum)

Figure 5.2   Formaldehyde (first band)

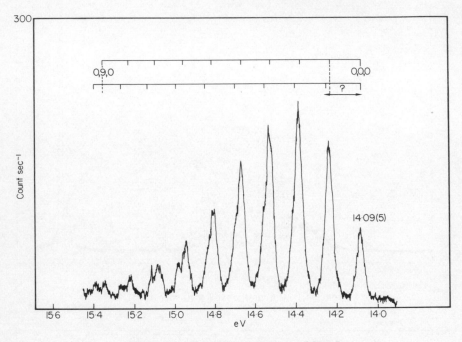

Figure 5.3   Formaldehyde (second band)

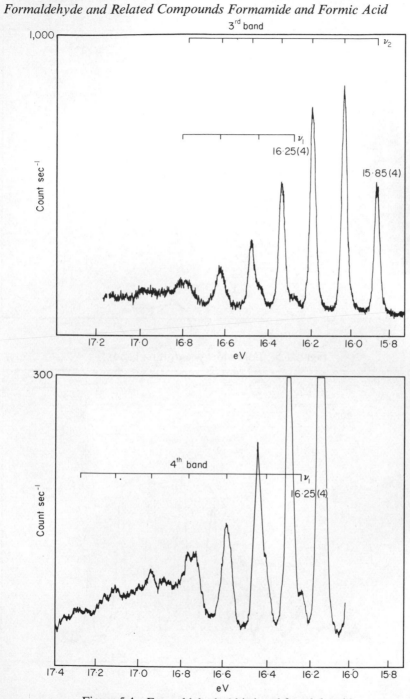

Figure 5.4 Formaldehyde (third and fourth bands)

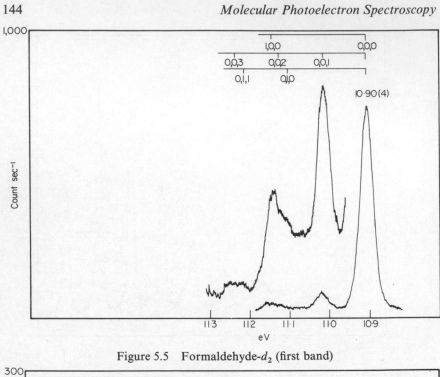

Figure 5.5   Formaldehyde-d_2 (first band)

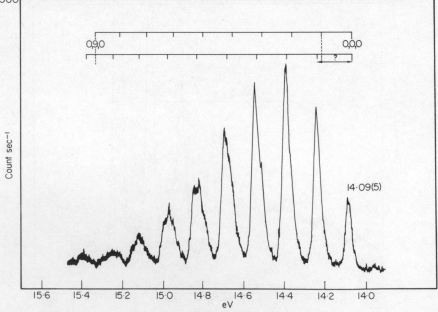

Figure 5.6   Formaldehyde-d_2 (second band)

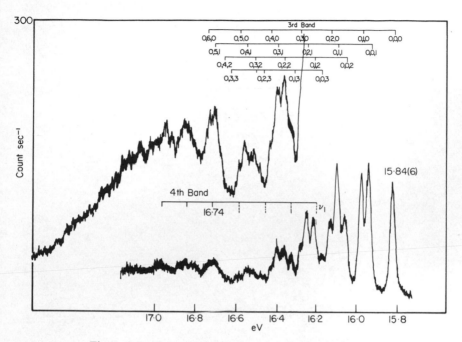

Figure 5.7   Formaldehyde-d_2 (third and fourth bands)

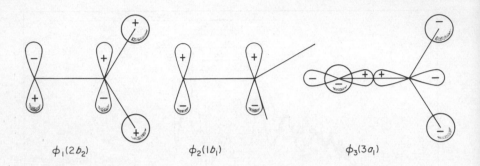

$\phi_1(2b_2)$ $\phi_2(1b_1)$ $\phi_3(3a_1)$

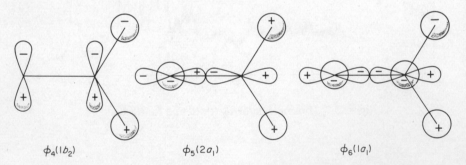

$\phi_4(1b_2)$ $\phi_5(2a_1)$ $\phi_6(1a_1)$

Figure 5.8   The occupied orbitals of the formaldehyde molecule illustrated in terms
of their constituent atomic orbitals

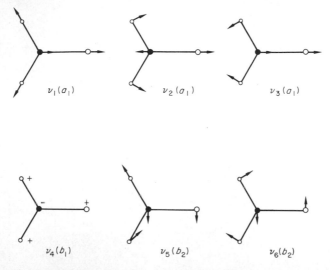

Figure 5.9   The normal modes of vibration for the form-aldehyde molecule

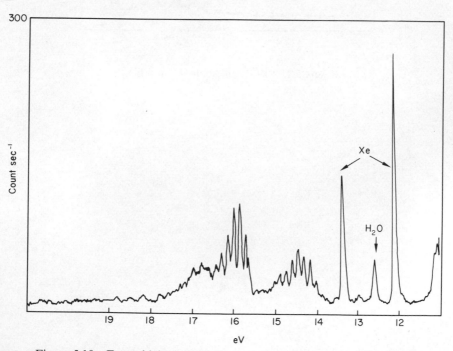

Figure 5.10   Formaldehyde-d_1 (full spectrum omitting part of first band)

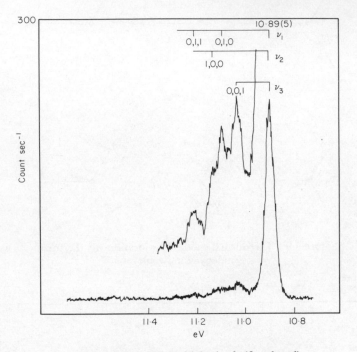

Figure 5.11   Formaldehyde-d_1 (first band)

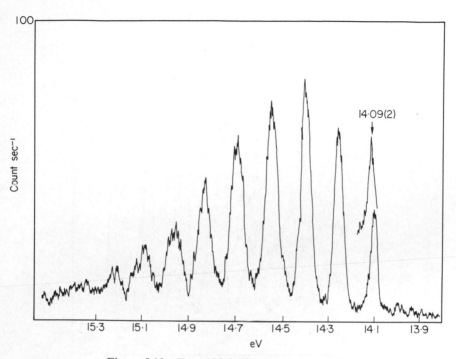

Figure 5.12   Formaldehyde-d_1 (second band)

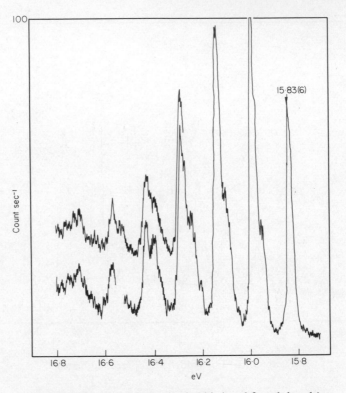

Figure 5.13   Formaldehyde-d_1 (third and fourth bands)

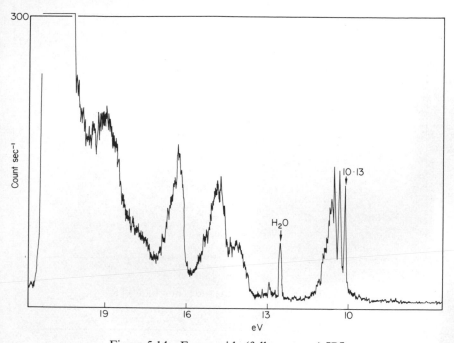
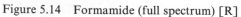

Figure 5.14   Formamide (full spectrum) [R]

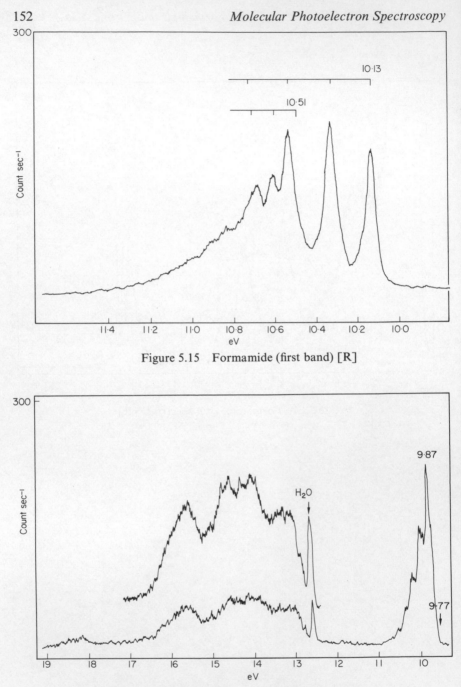

Figure 5.15   Formamide (first band) [R]

Figure 5.16   *N*-methylformamide (full spectrum)

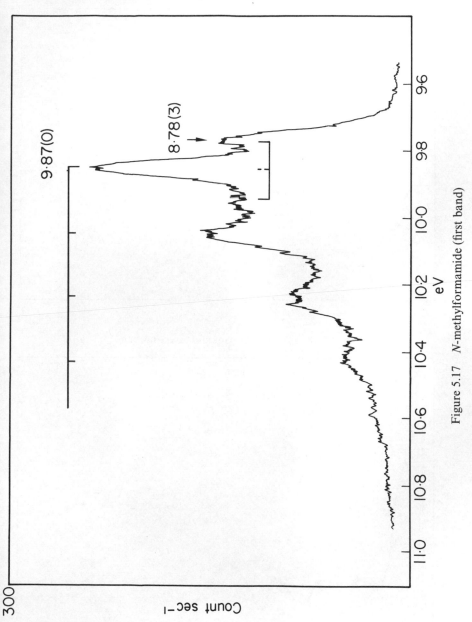

Figure 5.17 *N*-methylformamide (first band)

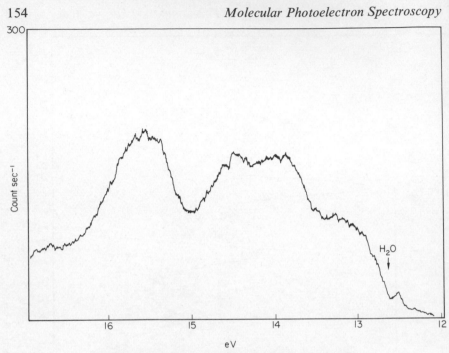

Figure 5.18   *N*-methylformamide (second band)

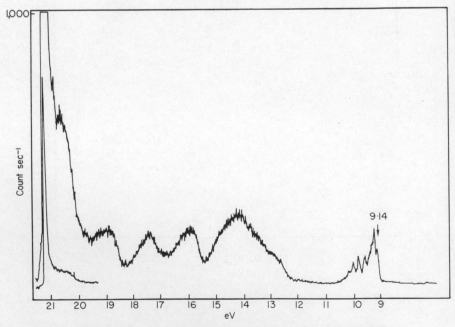

Figure 5.19   *N,N*-dimethylformamide (full spectrum) [R]

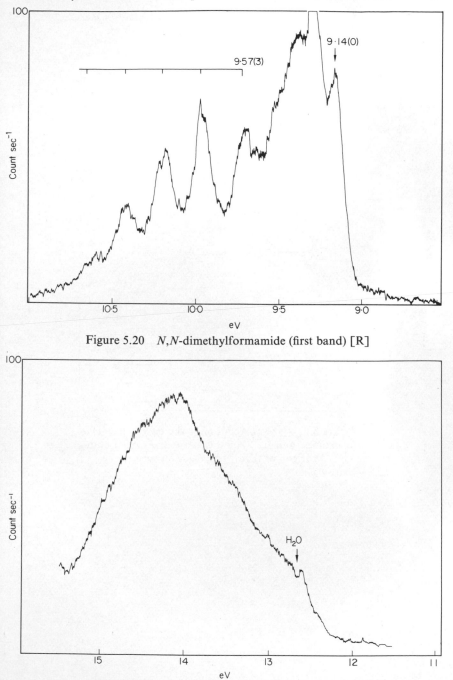

Figure 5.20   *N,N*-dimethylformamide (first band) [R]

Figure 5.21   *N,N*-dimethylformamide (second band) [R]

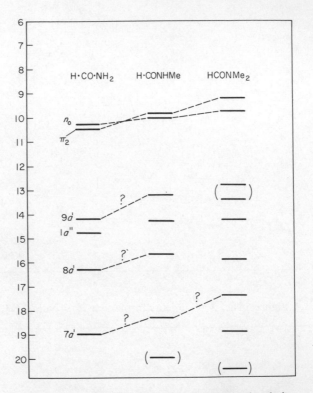

Figure 5.22 The correlation of the energy levels in formamide and its *N*-methyl derivatives from comparison of their photoelectron spectra

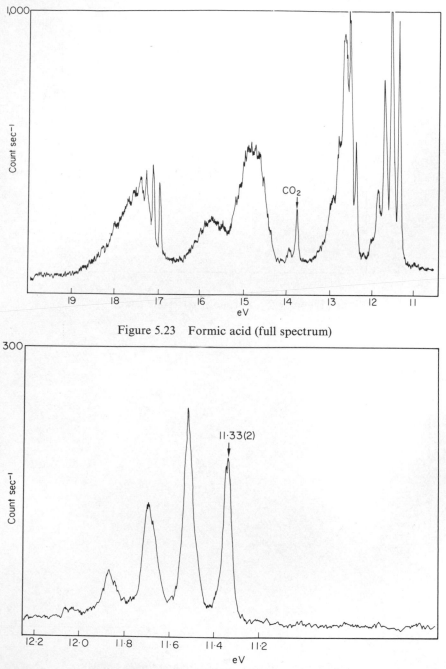

Figure 5.23   Formic acid (full spectrum)

Figure 5.24   Formic acid (first band)

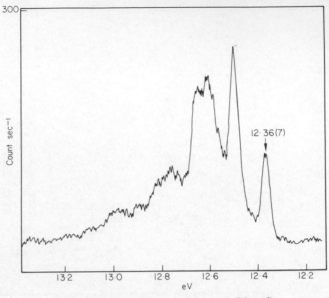

Figure 5.25   Formic acid (second band)

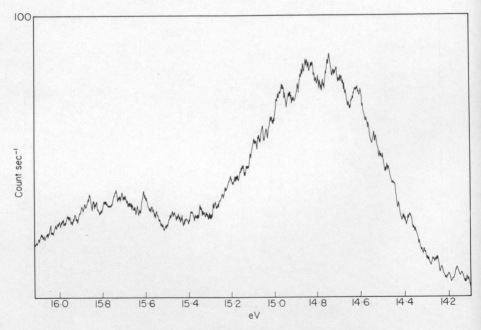

Figure 5.26   Formic acid (third and fourth bands)

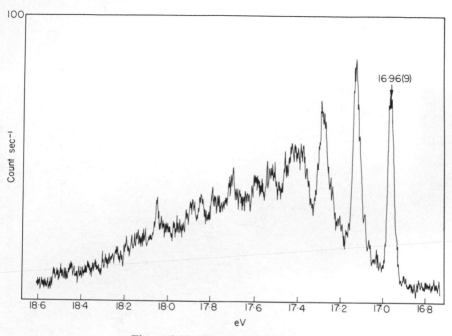

Figure 5.27   Formic acid (fifth band)

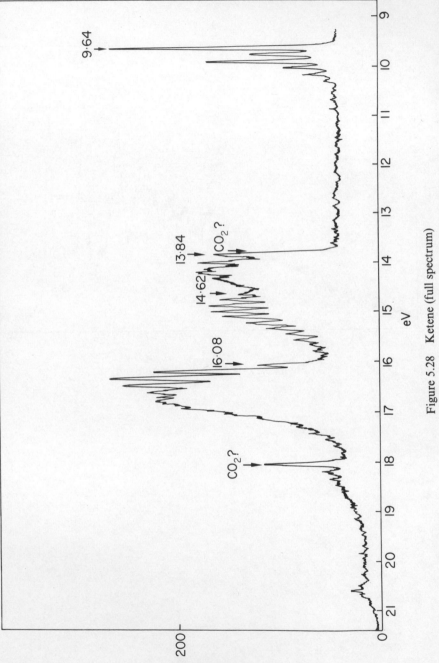

Figure 5.28   Ketene (full spectrum)

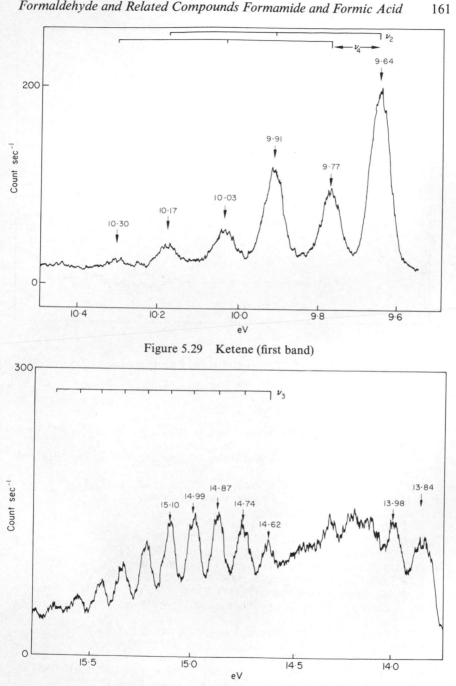

Figure 5.29 Ketene (first band)

Figure 5.30 Ketene (second and third bands)

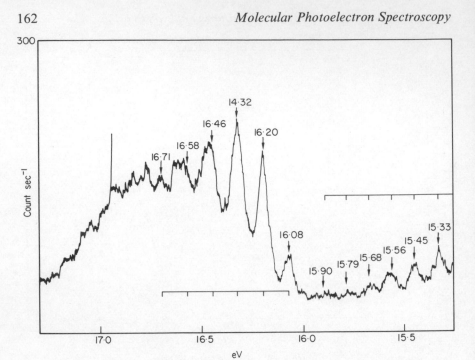

Figure 5.31   Ketene (third and fourth bands)

REFERENCES

1. A. D. Baker, C. Baker, C. R. Brundle and D. W. Turner, *Intern. J. Mass Spect. Ion Phys.*, **1**, 285 (1968).
2. G. Herzberg, *Molecular Spectra and Molecular Structure*, Vol. III, Van Nostrand, New York, 1966, p. 612.
3. T. M. Sugden and W. C. Price, *Trans. Faraday Soc.*, **44**, 116 (1948).
4. C. R. Brundle, *Ph.D. Thesis,* London University, 1968.
5. P. M. Guyon, *J. Chim. Phys.*, In the Press.
6. W. C. Price, *Chem. Rev.*, **41**, 257 (1947).
7. K. Higasi, I. Omura and H. Baba, *Nature*, **178**, 652 (1956).
8. C. R. Brundle and D. W. Turner, *Chem. Comm.*, **1967**, 314.
9. P. J. Goodfriend, F. W. Biss and A. B. Duncan, *Rev. Mod. Phys.*, **32**, 307 (1960).
10. J. M. Foster and F. M. Boys, *Rev. Mod. Phys.*, **32**, 303 (1960).
11. D. Peters, *Trans. Faraday Soc.*, **59**, 1121 (1963).
12. D. G. Carrol, L. G. Vanquickenborn and S. P. McGlynn, *J. Chem. Phys.*, **44**, 2779 (1966).
13. M. D. Newton and E. W. Palke, *J. Chem. Phys.*, **45**, 2329 (1966).
14. T. Shimanouchi and I. Suzuki, *J. Chem. Phys.*, **42**, 296 (1965).
15. D. P. May, *Ph.D. Thesis*, London University, 1966.
16. H. Basch, M. B. Robin and N. A. Keubler, *J. Chem. Phys.*, **47**, 1201 (1967).
17. H. Basch, M. B. Robin and N. A. Kuebler, *J. Chem. Phys.*, In the Press.

18. J. C. Evans, *J. Chem. Phys.*, **22**, 1228 (1954).
19. I. Suzaki, *Bull. Chem. Soc. Japan*, **33**, 1359 (1960).
20. P. G. Puranic, *Proc. Indian Acad. Sci. Sect. A*, **56**, 115 (1960).
21. C. C. Costain and J. M. Dowling, *J. Chem. Phys.*, **32**, 158 (1960).
22. M. B. Robin, Private communication.
23. K. Watanable, *J. Chem. Phys.*, **26**, 542 (1957).
24. J. D. Morrison and A. J. Nicholson, *J. Chem. Phys.*, **20**, 1021 (1952).
25. H. Pritchard and A. G. Harrison, *J. Chem. Phys.*, **48**, 2827 (1968).
26. G. Herzberg, *Molecular Spectra and Molecular Structure*, Vol. III, Van Nostrand, New York, 1966, p. 624.
27. W. C. Price, J. P. Teegan and A. D. Walsh, *J. Chem. Soc.*, **1951**, 920.
28. R. N. Dixon and G. H. Kirby, *Trans. Faraday Soc.*, **62**, 1406 (1966).

CHAPTER 6

Aliphatic Hydrocarbons

Values of ionization potentials of hydrocarbons are of paramount importance in theoretical chemistry, since they provide means for the evaluation of the often rather empirical procedures used for calculations on such molecules. Extensive use has been made of first ionization potential data in both empirical[1] and semiempirical[2] treatments of the electronic structure of hydrocarbons.

1. ALKANES

(A) Methane

For methane,[3] the electronic structure may be written as:

$$(1sa_1)^2 \ (2sa_1)^2 \ (2pt_2)^6 \ldots {}^1A_1$$

In this case, theoretical calculations consistently give the energy of the (pt_2) orbital in the region 12–15 eV,[4–11] whilst the energy of the $(2sa_1)$ orbital is usually placed between 24 and 25 eV.

Photoelectron spectroscopy at 584 Å (Figure 6.1) shows that methane has one ionization potential above -21.2 eV, the spectral band corresponding to this covering the range from 12·7 eV to about 16·0 eV, whilst photoelectron spectra obtained using a 304 Å ionizing source[12] and also using Al K_α radiation[13] reveal a second direct ionization process near 23 eV. In this instance the theoretical calculations are well substantiated by experiment.

The 12·7–16·0 eV band is very broad and can be seen to contain a double maximum. A number of vibrational peaks are also resolved on the edge of the band at low I.P.

Ionization from the highly bonding (pt_2) orbital would of course be expected to result in a large change in molecular dimensions, and the breadth of the photoelectron band is undoubtedly mainly due to this. To explain the double maximum in the band however it appears necessary to invoke the Jahn–Teller Theorem since this predicts that a tetrahedral triply degenerate methane ion cannot be a geometrically stable entity. Thus the photoelectron spectrum may well show transitions from the methane molecular ground state to ions of lower symmetry than T_d;

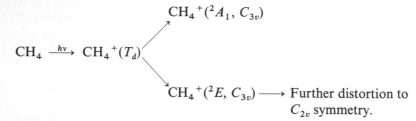

$$CH_4 \xrightarrow{h\nu} CH_4{}^+(T_d)$$

$$CH_4{}^+(^2A_1, C_{3v})$$

$$CH_4{}^+(^2E, C_{3v}) \longrightarrow \text{Further distortion to } C_{2v} \text{ symmetry.}$$

The fine structure resolved in the photoelectron spectrum of methane admits of no simple analysis—again this may be ascribed to Jahn–Teller effects, since these would give a large density of irregularly spaced vibrational levels.

(B) Ethane

The 'putting together' of two methyl groups to give ethane has the effect of lifting the degeneracy of the (pt_2) orbital, giving rise to three orbitals, (e_g), (e_u) and (a_{1g}). In addition, the (sa_1) orbital of methane splits to give (a_{1g}) and (a_{2u}) orbitals for ethane.

Calculations of the orbital energies[11] gives the order of the orbitals and allows the full electronic structure of ethane to be written as:

$$(1a_{2u})^2 \, (^1a_{1g})^2 \, (2a_{1g})^2 \, (2a_{2u})^2 \, (1e_u)^4 \, (3a_{1g})^2 \, (1e_g)^4 \ldots {}^1A_1$$

The $(1e_g)$ and $(3a_{1g})$ orbitals are calculated to have very similar energies, and in the photoelectron spectrum[3] (Figures 6.2–6.4) the two relevant bands (vertical I.P.'s ~ 12 and 13 eV respectively) overlap considerably. In fact the ethane ion should be susceptible to distortion by Jahn–Teller forces following ionization from the $(1e_g)$ orbital and so the 11·5–14·5 eV region of the spectrum may consist of two bands relating to electron loss from the highest occupied orbital (cf. methane) overlapping with the band associated with ionization from the $(3a_{1g})$ orbital.

Vacuum ultraviolet absorption studies have generally afforded no vibrational fine structure, and this was explained[14] in terms of the instability of all upper states. Recently however, other workers have found vibrational structure in Rydberg levels leading to the first I.P. of ethane[15,16], and photoionization studies[17] have detected vibrational levels on the first ionization. The photoelectron spectrum (Figure 6.3) shows a well resolved series of peaks the spacing of which being 1170 cm^{-1}. This is interpretable in terms of either the CH_3 deformation (ν_2)[15], or the C—C stretching mode (ν_3).[16]

The excited state of the ethane ion formed by electron loss from the degenerate $(1e_u)$ orbital should also be 'Jahn–Teller unstable'. The corresponding band in the spectrum, 14·7–16·5 eV, shows slight evidence of a division into two parts.

The ($2a_{2u}$) orbital has some C—C antibonding character[11], and the vibrational frequency observed in the 20·13 eV band (Figure 6.4), 1160 cm^{-1}, is probably the C—C stretching mode v_3. In hexadeutero ethane, this interval is reduced to 1070 cm^{-1}, the decrease being almost exactly that expected to occur in the C—C stretching vibration upon deuteration.

(C) Higher Paraffins

The orbital energies of the paraffins have been calculated by Hoffmann[9], using the LCAO–MO method. The effect whereby his calculated bonding levels fall into two sections is observed experimentally[18], although Hoffman's results are consistently higher than the observed I.P.'s. The I.P.'s of some simple hydrocarbons are collected in Figure 6.5. The Franck–Condon envelope is indicated in the case of methane. The results for hydrocarbons higher than ethane have been obtained using the low resolution magnetic spectrometer described in Chapter 2.

2. ALKENES

(A) Ethylene

The photoelectron spectrum of ethylene[3] (Figures 6.6–6.9) shows that there are five occupied orbitals with energies above −21·2 eV, the adiabatic I.P.'s being 10·51, 12·38, 14·4(7), 15·6(8) and 18·8(7) eV. Only for the first two can a reasonably accurate estimate be given, since the 0←0 components of the other three bands are not clearly resolved.

Ethylene is the simplest organic molecule to contain a single C—C π bond, and it is ionization from this π orbital which is responsible for the lowest ionization band in the spectrum. The envelope of vibrational peaks associated with this band (i.e. a strong 0←0 component followed by a short series of weaker peaks) is found to be fairly characteristic of π ionization. Expansion of the band (Figure 6.6) reveals that what at first sight appears to be a simple series of peaks is in fact a series of doublets. This is readily explicable however if we assume that the totally symmetric C=C valence stretching mode (v_2) is excited in conjunction with the twisting vibration (v_4).

Each of the photoelectron bands shows complex fine structure, which we have attempted to analyse in the light of the known vibrational modes of the ethylene molecule (see Figure 6.10 and Table 6.1). Some of these vibrational analyses and assignments must be regarded as speculative since the fine structure is only partially resolved in each band. Furthermore, in the third band (14·47 eV), the great breadth of the vibrational peaks rather implies that the

Table 6.1   Vibrational modes and frequencies identified from the photoelectron spectrum of ethylene and the corresponding values in the molecule

State	Vibrations
$C_2H_4(X)$ $(^1A_{1g})$	ν_1 Totally symmetric C—H stretch, 3026 cm^{-1} ν_2 Totally symmetric C—C valence stretch, 1623 cm^{-1} ν_3 C—H deformation, 1342 cm^{-1} ν_4 Twisting vibration, 1027 cm^{-1}
$C_2H_4^+(\tilde{X})$	$\nu_2 = 1230 \pm 50$ cm^{-1} $\nu_4 = 430 \pm 50$ cm^{-1} $\nu_3(?) \simeq 1340$ cm^{-1}*
$C_2H_4^+(\tilde{A})$	$\nu_1 = 2900 \pm 50$ cm^{-1}* $\nu_2 = 1300 \pm 50$ cm^{-1}* $\nu_3 = 800 \pm 50$ cm^{-1}*
$C_2H_4^+(\tilde{B})$	$\nu_1 \simeq 1700$ cm^{-1}
$C_2H_4^+(\tilde{C})$	ν_1 or $\nu_3 \simeq 1240$ cm^{-1}

* Indicates speculative values or assignments

vibrational structure involves something more than a simple progression in ν_1—the C—H stretching mode. However, the indication is that in this case the main dimensional change on the third ionization is in the C—H bond. This is interesting, since in comparison with the majority of sophisticated LCAO–SCF calculations employing both Slater and Gaussian basis sets (Table 6.2) all except Berthod's results[19] indicate that the third ionization will involve the $(3a_g)$ orbital (both C—H *and* C—C bonding). Again, our results seem to show that both the C—C and C—H vibrational modes are involved in the second ionization, whereas the majority of the calculations have $(1b_{1g})$ (mostly C—H bonding) as the second highest occupied orbital. A reversal of the ordering of the $(3a_g)$ and $(1b_{1g})$ orbitals is therefore indicated, since

Table 6.2   Calculated orbital energies (eV) of ethylene

Ref.	$1a_g$	$1b_{3u}$	$2a_g$	$2b_{3u}$	$1b_{2u}$	$3a_g$	$1b_{1g}$	$1b_{1u}$
11	307	307	27·6	21·3	17·5	15·3	13·8	10·1
19	311	310	24·5	20·7	20·6	15·1	18·9	8·8
20	307	307	27·5	21·3	17·5	15·3	13·8	10·0
21	306	306	28·8	22·0	18·0	15·9	14·1	10·4
22	306	306	28·1	21·7	17·6	15·9	13·8	10·0
23	306	306	28·1	21·7	17·6	15·9	13·8	10·0
24	296	296	28·6	21·4	17·6	15·4	13·3	10·5

this would reconcile the vibrational structure observed in the photoelectron bands with the bonding characteristics of the orbitals involved.

(B) Butadiene

If we consider the butadiene molecule to arise from two ethylenic systems, with the double bonds conjugated, the two ethylene π orbitals, only one of which is occupied in the ground state, interact to form four new π orbitals, two of which are occupied in the ground state. It is possible for the molecule to exist in *s-cis* and *s-trans* forms, for which the selection rules for electronic transitions will be different.[25] In the *s-cis* case, all transitions are allowed, but for the *s-trans* molecule, $N \rightarrow V_2$ and $N \rightarrow V_3$ are forbidden. However the $N \rightarrow V_2$ transition occurs at room temperature with sufficient intensity to indicate the presence of some of the *s-cis* form. Calorimetric studies[26] have indicated the presence of more than 1% of the *s-cis* form, while the use of chemical techniques[27] (the analysis of 1,4-addition compounds) has suggested that as much as 3–7% may be present.

The ionization potential of *trans*-butadiene has been found spectroscopically to be 9·062 eV.[28] In addition, Sugden and Walsh[29] have reported an ionization potential about 0·3 eV lower than this, on the evidence of a break in the electron impact efficiency curve and weak bands in the u.v. spectrum which could be arranged in a Rydberg series. However, this lower I.P. was not detected by photoionization[30, 31] or in earlier photoelectron spectroscopic[32, 33] studies. In fact, calculations predict the I.P. of the *s-cis* form to be greater than that of the *s-trans* form by 0·03–0·14 eV.[32] The weak bands at about 8·6 and 11·0 eV are due to an impurity since in the spectrum of a second sample they were not present.

There is general agreement as to the value of the ionization potential of butadiene (9·07 eV, [30, 32, 34] 9·075 eV, [31] 9·08 eV[33]), and the high resolution photoelectron spectrum (Figure 6.11) shows that the orbital involved is rather nonbonding, and is therefore likely to be the upper π orbital ($1b_g$). It is not possible to decide, *a priori*, which of the 11·4 or 12·2 eV bands is associated with the lower π orbital ($1a_u$). Calculations reported recently by Dewar[35] place the energy of the highest occupied σ orbital between those of the two π orbitals, while Kato and coworkers[36] assign the two orbitals of highest energies as being π.

The first band in the photoelectron spectrum (Figure 6.12), at adiabatic I.P. 9·08 eV, indicates that two, and possibly three, vibrational modes are excited in the ionic ground state, the frequencies observed being 1520, 1180 and 500 cm^{-1}. The first of these can be assigned to the symmetric C=C stretching mode, which has a ground state frequency of 1643 cm^{-1}, and the last, to the skeletal bending mode, which has a frequency of 513 cm^{-1} in the molecule. The other mode excited is expected to be the C—C stretching vibration, and to this is assigned the 1180 cm^{-1} frequency. This mode is reported by Harris[37] to have a molecular frequency of 1205 cm^{-1}, but since the orbital is C=C bonding but C—C antibonding, the ionic frequency of the C—C stretching mode would be expected to be higher than in the molecule. Indeed, calculations[38] of the force constants and bond lengths in butadiene and the butadiene ion indicate an increase in the C=C bond lengths, but a decrease in that of the C—C bond. It may be that the assignments[37] of the C—C stretching and the methylene C—H rocking modes should be reversed, as appears to be the case also for acrolein (Chapter 9). This would assign the C—C stretching mode a molecular frequency of about 900 cm^{-1}, more in keeping with the data obtained from the photoelectron spectrum.

The ionization energies of butadiene are compared with those of acrolein and glyoxal, with which it is isoelectronic, in Chapter 9.

(C) Allene, Methylene *Cyclo*propane, Methylene *Cyclo*butane and Methylene *Cyclo*pentane

In this short series of compounds, one of the major structural variations is the change in the angle of the bonds on one of the unsaturated carbon atoms. The effect that this has on the bonding properties of the C—C π orbital is reflected in the lowering of the π I.P., and also in the different Franck–Condon envelopes obtained for the π band in the photoelectron spectra. We infer for example that allene and methylene cyclopentane (Figures 6.13 and 6.18) have a π orbital appreciably more bonding in the double bond than that in methylene cyclobutane (Figure 6.16).

The π band in the spectra of the cyclobutane and cyclopentane compounds indicates the excitation of two vibrational frequencies in the ground states of the ions (Figures 6.17 and 6.19). One of these is almost certainly the C=C stretching mode ($C_5H_8^+(\tilde{X})$ 1320 cm^{-1}, $C_6H_{10}^+(\tilde{X})$ 1120 cm^{-1}). These are somewhat lower than the frequencies observed in the molecular ground states ($C_5H_8(X)$ 1678 cm^{-1}, $C_6H_{10}(X)$ 1657 cm^{-1}),[39] that in the methylene cyclopentane ion being reduced more than that in the methylene cyclobutane ion in accordance with the more bonding properties of the π orbital.

The first band in the allene spectrum (Figure 6.14) gives evidence of Jahn–Teller instability in the ground state of the $C_3H_4^+$ ion. Thus there are two transitions with vertical I.P.'s $\sim 10\cdot0$ and $\sim 10\cdot6$ eV, and also there is irregular spacing of the vibrational fine structure. This fine structure may indicate the excitation of a vibration with frequency about 700 cm^{-1}. This is probably the C=C=C stretching mode v_3. This mode has been observed in the ultraviolet spectrum with a frequency of 610 ± 100 cm^{-1}.[26]

3. ALKYNES

(A) Acetylene

The photoelectron spectra of compounds containing acetylenic linkages all show prominent bands characteristic of the loss of an electron from π orbitals. These bands are easily recognized by the shapes of the Franck–Condon envelopes. Thus in the acetylene spectrum (Figures 6.20–6.22) it is clear that the first band, adiabatic I.P. 11·40 eV, corresponds to the ejection of an electron from the least strongly bonding molecular orbital. This is consistent with the accepted electronic structure of the ground state of the acetylene molecule, being:

$$(\sigma_g 1s_c)^2 \; (\sigma_u 1s_c)^2 \; (2\sigma_g)^2 \; (2\sigma_u)^2 \; (3\sigma_g)^2 \; (1\pi_u)^4 \ldots {}^1\textstyle\sum_g{}^+$$

The shape of the first band, namely a short series of well resolved peaks (Figure 6.21) is characteristic of ionization from the highest occupied π orbital in all the alkynes studied. Electron removal from the $^1\pi_u$ orbital of acetylene predictably excites the C≡C stretching vibration (ν_2), the frequency in the ion being only a little lower than that in the molecular ground state (Table 6.3) again indicating weak bonding character.

Table 6.3   Vibrational modes and frequencies identified from the photoelectron spectrum of alkynes and the corresponding values in the molecule

Molecule	Ionic state	I.P. (eV)	Vibrational mode excited	Ionic frequency (cm^{-1})	Molecular frequency (cm^{-1})	Description of mode
C_2H_2	$(\tilde{X})$	11·40	ν_2	1830	1983	C≡C stretching
	$(\tilde{A})$	16·36				
	$(\tilde{B})$	18·38	ν_1	1900	3369	Symm. C—H stretching
			$\nu_2(?)$	2510		
C_2D_2	$(\tilde{X})$	11·40	ν_2	1610	1761	
	$(\tilde{A})$	16·53	ν_1	2500	2683	
			ν_2	1370(?)		
	$(\tilde{B})$	18·44	ν_1	1420		
			ν_2	2290		
C_4H_2	$(\tilde{X})$	10·17	ν_2	2120	2184	Symm. C≡C stretching
			ν_4	2820	3329	Asymm. C—H stretching
	$(\tilde{A})$	12·62	ν_2	1860		
			ν_3	810	874	C—C stretching
	$(\tilde{B})$	16·61				
	$(\tilde{C})$	19·8				

Table 6.3—*continued*

Molecule	Ionic state	I.P. (eV)	Vibrational mode excited	Ionic frequency (cm^{-1})	Molecular frequency (cm^{-1})	Description of mode
C_4D_2	$(\tilde{X})$	10·18	ν_2	1980		
			ν_4	2180		
	$(\tilde{A})$	12·62	ν_2	1770		
			ν_3	800		
	$(\tilde{B})$	16·74				
	$(\tilde{C})$	19·8				
C_3HN	$(\tilde{X})$	11·60	ν_2	2180	2271	C≡C & C≡N in-phase stretching
	$(\tilde{A})$	13·54	ν_4	860	876	C—C stretching
	$(\tilde{B})$	14·03	ν_2	1940		
			ν_4	810		
	$(\tilde{C})$	17·62	ν_1	1320	3329	C—H stretching
C_4N_2	$(\tilde{X})$	11·81	$\nu_2(?)$	2200	2290	Symm. C≡N stretching
	$(\tilde{A})$	13·89				
	$(\tilde{D})?$	14·95	ν_2	2100	2119	C≡C stretching
			ν_3	590	692	Symm. C—C stretching
C_3H_4	$(\tilde{X})$	10·37	ν_3	1940	2142	C≡C stretching
			ν_5	940	930	C—C stretching
	$(\tilde{A})$	13·69	ν_4	1290	1382	CH$_3$ deformation
	$(\tilde{B})$	15·2				
	$(\tilde{C})$	17·2				
C_8H_6	$(\tilde{X})$	8·60		2180		
				1090		
	$(\tilde{A})$	10·63		2020		
				970		
	$(\tilde{B})$	12·10		850		
				300		
	$(\tilde{C})$	13·77				
	$(\tilde{D})$	15·1				
	$(\tilde{E})$	17·52				
	$(\tilde{F})$	19·3				

In general, the appearance of vibrational fine structure is not restricted to π bands, though in these cases it seems easier to interpret than when associated with σ bands. For instance, the second band in the acetylene spectrum

(Figure 6.21) clearly contains structure but it is not possible to assign it completely in terms of regular vibrational series. The effect of deuteration on the vibrations excited however results in simpler structure in the second band of the C_2D_2 spectrum (Figure 6.24), interpretable as excitation of the $C{\equiv}C$ (v_2) and the C—D (v_1) stretching modes.

The third band in the C_2H_2 spectrum (Figure 6.22) and the C_2D_2 spectrum (Figure 6.25) can also be analysed in terms of the $C{\equiv}C$ and the C—H (or C—D) stretching modes. The C—H vibrational frequency is much lower than is observed in the molecule, owing to the bonding nature of the orbital. The $C{\equiv}C$ vibrational frequency is higher than the molecular frequency however. This is in accord with descriptions[11] of the corresponding orbital as being C—H bonding and C—C antibonding.

(B) Diacetylene

The effect of a second carbon–carbon 'triple bond' in diacetylene, in terms of simple Hückel theory, is to generate from the acetylene $^1\pi_u$ orbitals of two isolated molecules two new molecular orbitals, $^1\pi_u$ and $^1\pi_g$, the energies of which are symmetrically above and below that of the original acetylene $^1\pi_u$ orbital (Figure 6.26). The higher of the two π orbitals, I.P. 10·17 eV, gives the characteristic band shape, and predictably, the vibration excited is the symmetric $C{\equiv}C$ stretching mode $(v_2)^{43}$.

The other occupied π orbital gives a differently shaped band in the spectrum, I.P. 12·62 eV. This Franck–Condon envelope is found however to be characteristic of bands obtained from electrons ejected from the lowest-occupied π orbitals of this type. The more complex fine structure can be interpreted in terms of the excitation of two vibrational modes, viz. the C—C stretching (v_3) and the symmetric $C{\equiv}C$ stretching (v_3) and the symmetric $C{\equiv}C$ stretching modes (v_2) both lower than those occurring in the molecule.

(C) Cyanoacetylene

Cyanoacetylene, which is isoelectronic with diacetylene, gives rise to a photoelectron spectrum in which the two π bands can easily be picked out at adiabatic I.P.'s 11·60 and 14·03 eV (Figures 6.27, 6.28). The vibrations excited in these two bands are those which would be expected (Table 6.3). The second band in the spectrum, with its intense $0{\leftarrow}0$ component occurring at 13·54 eV, indicates that the ionizing transition forming the first excited ionic state is from an almost completely nonbonding orbital. It is therefore correlated with the nitrogen $2p\sigma$ orbital which gives a similar single, sharp peak.

The highest ionization energy band, at 17·62 eV, is very similar in form to that described in the hydrogen cyanide spectrum (Chapter 13). It can be related to an orbital which is essentially C—H σ bonding, the vibrational mode excited being best described as the C—H stretching mode greatly reduced, compared with that in the molecular ground state, because of the strong bonding nature of the orbital. The continuum which appears at about 18·5 eV in this band is probably a result of dissociation of the C_3HN^+ ion, at these higher energies, to C_3N^+, which has an appearance potential at about $18\cdot2 \pm 0\cdot3$ eV.[27]

The replacement of a C—H group in diacetylene by a nitrogen atom to give cyanoacetylene has the effect of raising the ionization potentials of the π orbitals, owing to the increased electronegativity of the hetero atom compared with carbon. This effect is enhanced by the inclusion of a second nitrogen atom, giving cyanogen (Chapter 13).

(D) Dicyanoacetylene

In dicyanoacetylene there are competing effects of the raising of the I.P.'s by the more electronegative nitrogen atoms and their lowering by conjugation. The result is that the π orbitals of dicyanoacetylene have similar energies to those of cyanoacetylene. The bands at 11·81 and 14·95 eV in the photoelectron spectrum (Figures 6.29–6.31), can readily be identified with the $(2\pi_u)$ and $(1\pi_u)$ orbitals respectively. The second band, at 13·89 eV, appears at an energy appropriate to nitrogen $(2p + 2p)$ and $(2p - 2p)$ σ orbitals. The complexity of the structure of the band and the integrated areas suggest that it may in fact be due to overlapping of the three expected bands, nitrogen $(2p + 2p)$, nitrogen $(2p - 2p)$ and $(^1\pi_g)$, the 'middle' π level.

(E) Methyl Acetylene and Dimethyl Triacetylene

The fine structure associated with the π bands of methyl substituted acetylenes is slightly more complex than for the corresponding parent compounds, owing to the excitation of the C—C stretching mode in addition to the C≡C stretching mode. This is a result of overlap of the carbon–carbon π orbitals with the methyl 'pseudo π' orbitals, producing π orbitals which are delocalized over the whole carbon skeleton. Thus the π band in the methyl acetylene spectrum (Figure 6.32), at 10·37 eV, indicates excitation of the C—C and C≡C stretching modes. The π bands of the dimethyl triacetylene spectrum (Figure 6.33) may also be interpretable in these terms.

The ionization potentials of the alkynes described in this chapter, and vibrational modes excited upon ionization, are summarized in Table 6.3. A discussion of the critical potentials of acetylene, which have been obtained by various techniques, has been given by Collin and Delwiche,[28] and the relation between the ionization potentials of the two series of isoelectronic molecules, nitrogen, hydrogen cyanide, and acetylene, and also diacetylene, cyanoacetylene and cyanogen has been discussed by Baker and Turner.[29]

The ionization potentials of the molecules mentioned in this section, to-gether with those of cyanogen, are compared in the form of an energy level diagram (Figure 6.34).

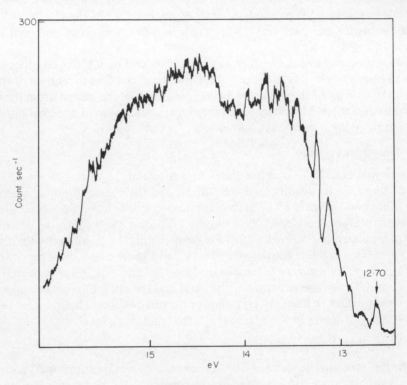

Figure 6.1   Methane

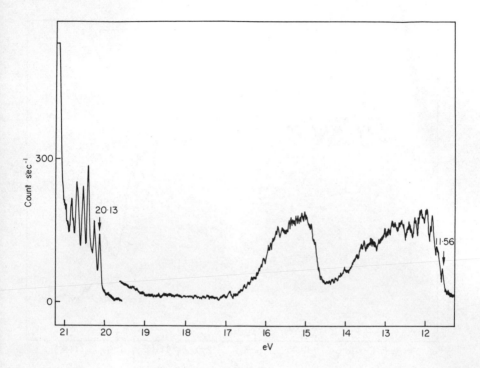

Figure 6.2   Ethane (full spectrum) [R]

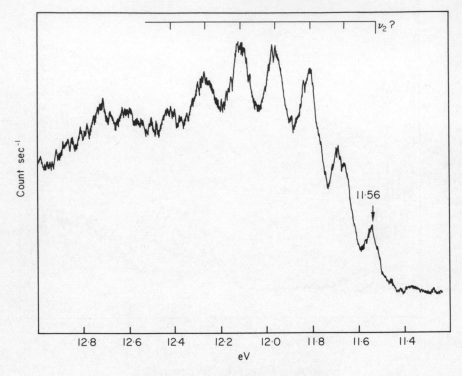

Figure 6.3   Ethane (first band) [R]

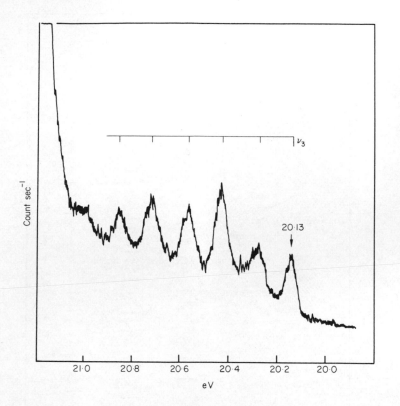

Figure 6.4   Ethane (third band) [R]

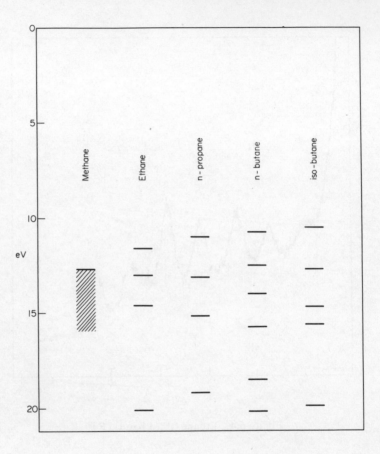

Figure 6.5   Energy levels of some saturated paraffin molecules
inferred from their photoelectron spectra

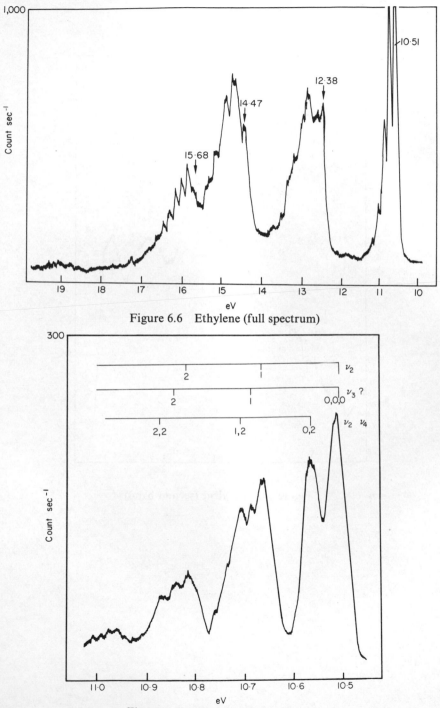

Figure 6.6 Ethylene (full spectrum)

Figure 6.7 Ethylene (first band)

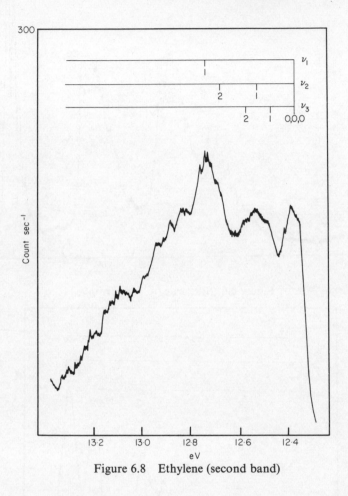

Figure 6.8   Ethylene (second band)

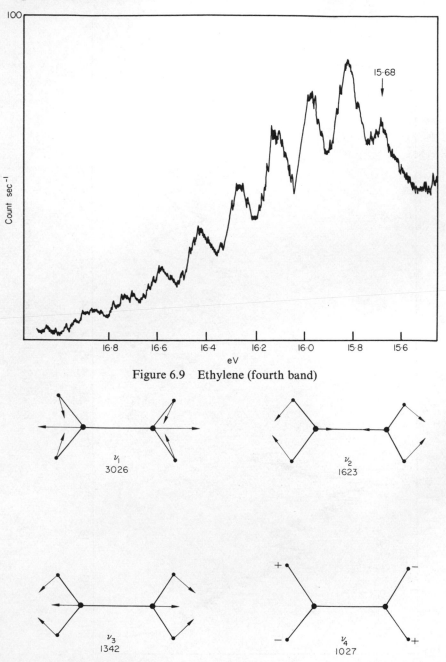

Figure 6.9 Ethylene (fourth band)

ν_1
3026

ν_2
1623

ν_3
1342

ν_4
1027

Figure 6.10 The symmetric normal modes of vibration of the ethylene molecule

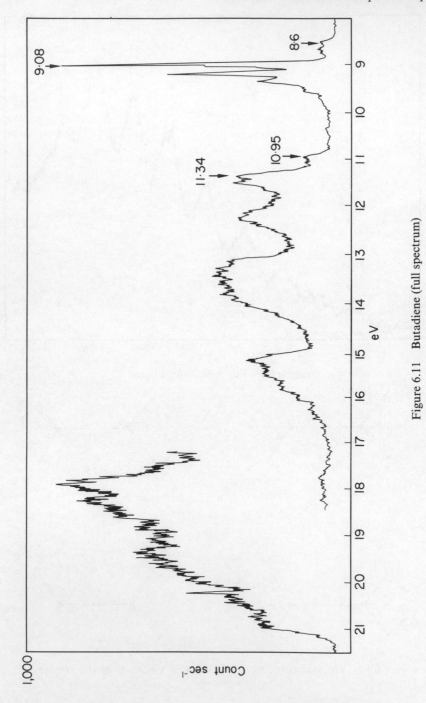

Figure 6.11 Butadiene (full spectrum)

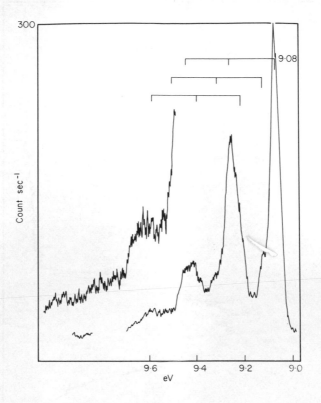

Figure 6.12   Butadiene (first band)

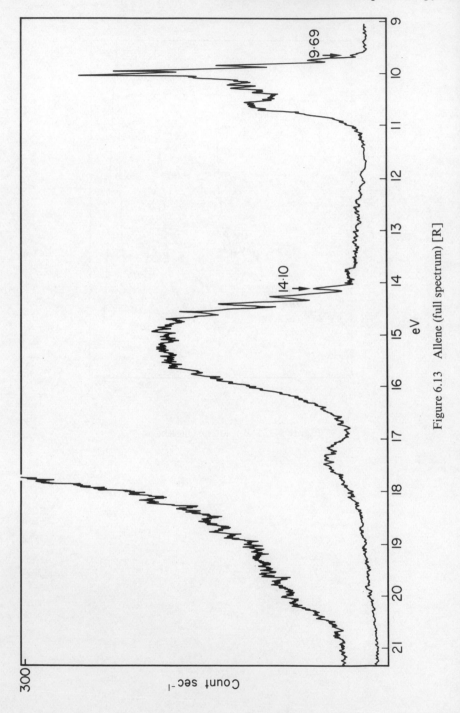

Figure 6.13 Allene (full spectrum) [R]

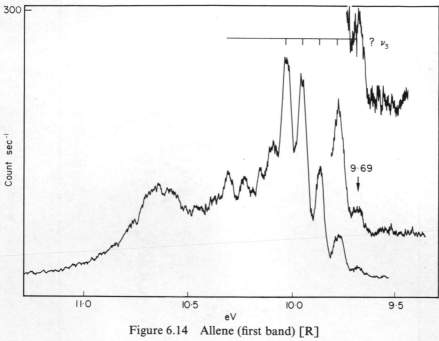

Figure 6.14   Allene (first band) [R]

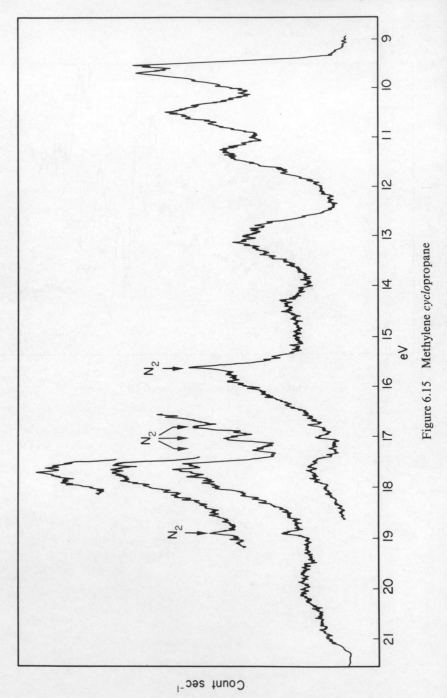

Figure 6.15 Methylene *cyclo*propane

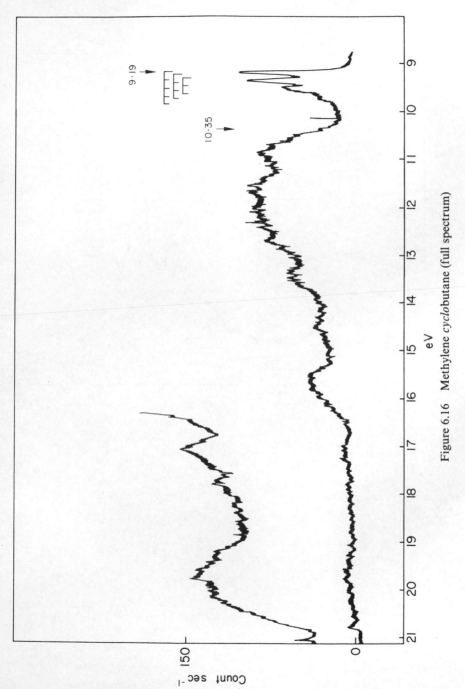

Figure 6.16 Methylene *cyclobutane* (full spectrum)

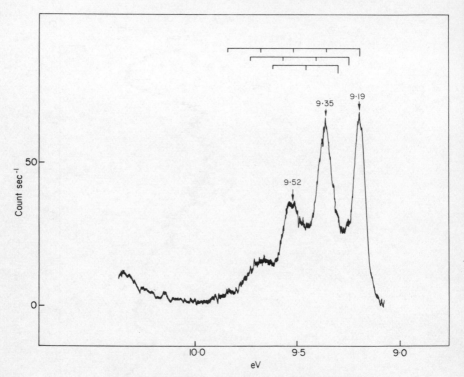

Figure 6.17   Methylene *cyclo*butane (first band)

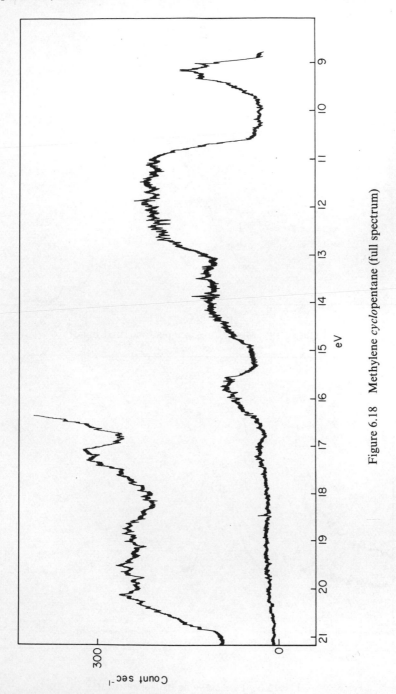

Figure 6.18   Methylene *cyclopentane* (full spectrum)

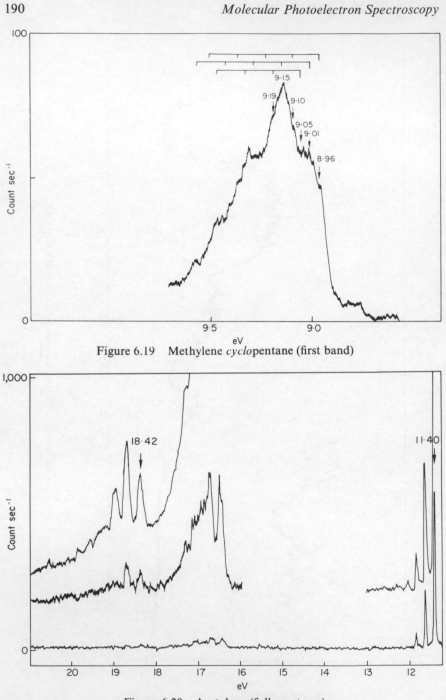

Figure 6.19   Methylene *cyclo*pentane (first band)

Figure 6.20   Acetylene (full spectrum)

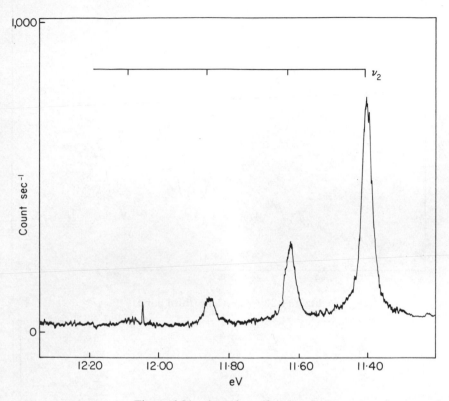

Figure 6.21   Acetylene (first band)

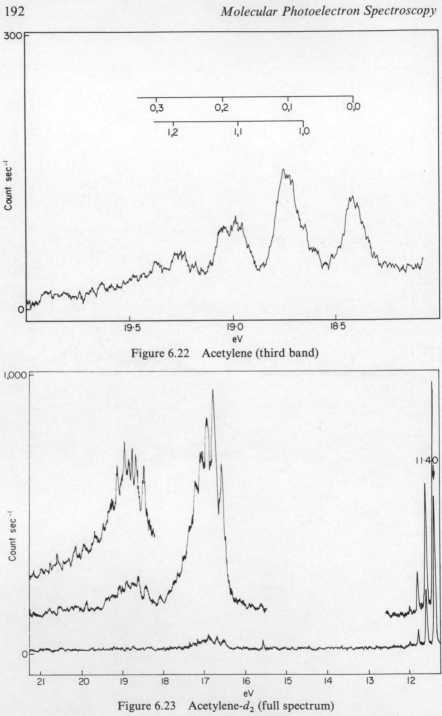

Figure 6.22   Acetylene (third band)

Figure 6.23   Acetylene-d_2 (full spectrum)

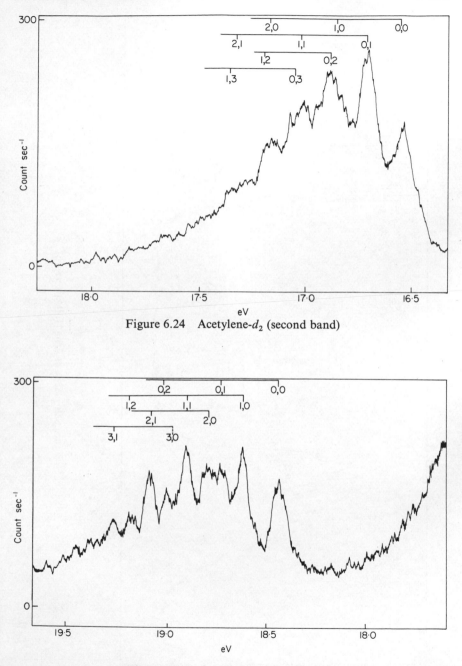

Figure 6.24   Acetylene-d_2 (second band)

Figure 6.25   Acetylene-d_2 (third band)

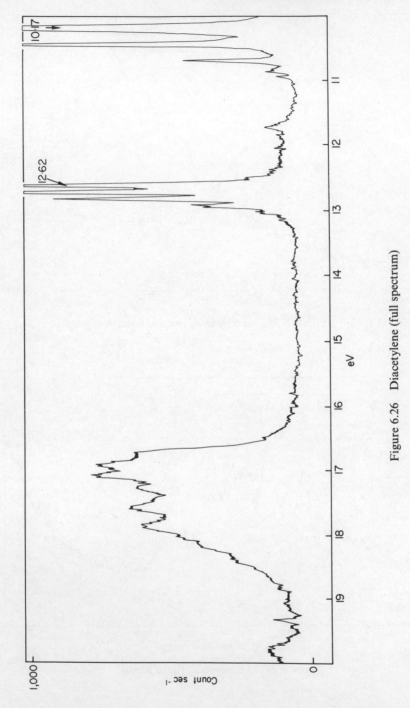

Figure 6.26   Diacetylene (full spectrum)

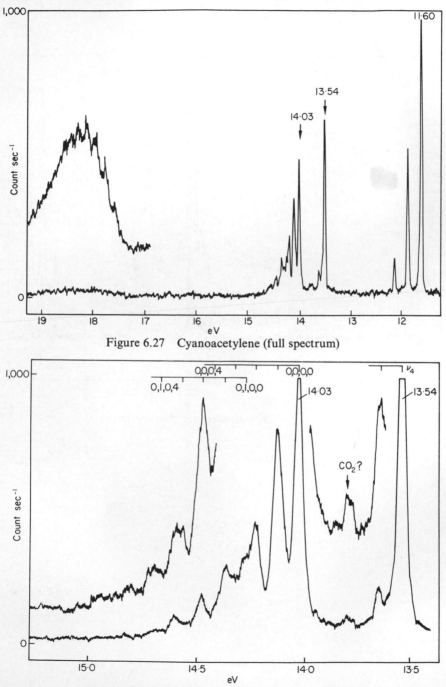

Figure 6.27 Cyanoacetylene (full spectrum)

Figure 6.28 Cyanoacetylene (second and third bands)

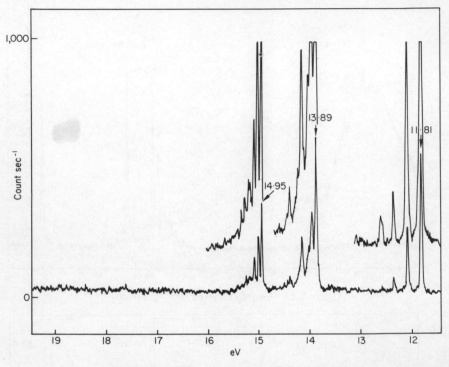

Figure 6.29 Dicyanoacetylene (full spectrum)

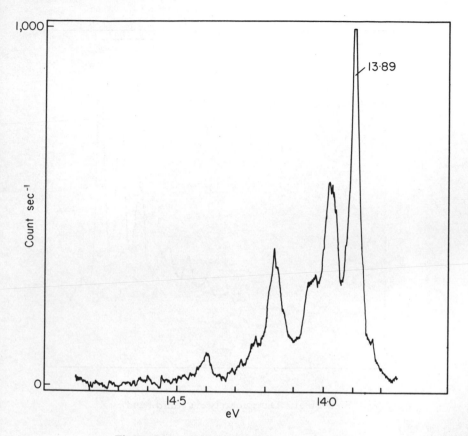

Figure 6.30   Dicyanoacetylene (second band)

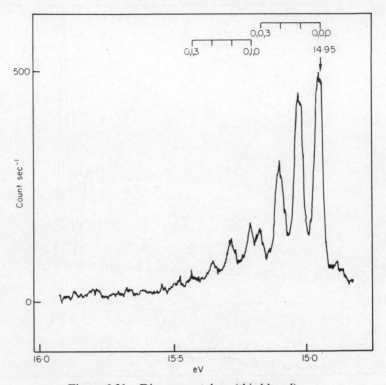

Figure 6.31   Dicyanoacetylene (third band)

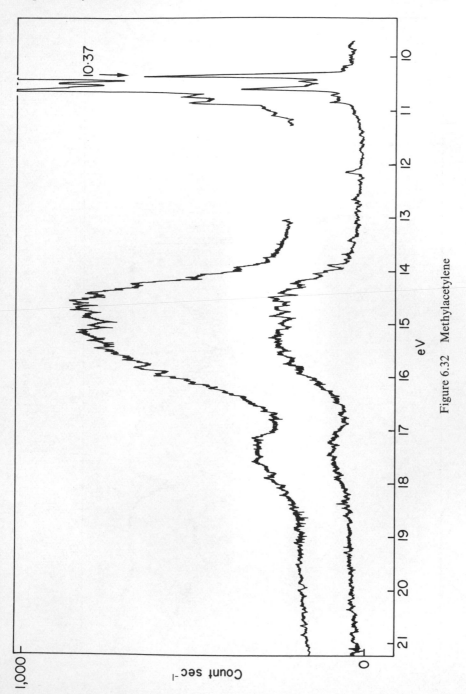

Figure 6.32 Methylacetylene

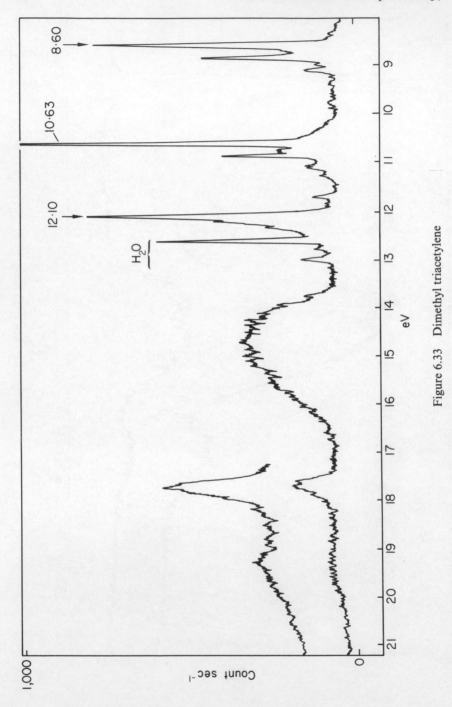

Figure 6.33   Dimethyl triacetylene

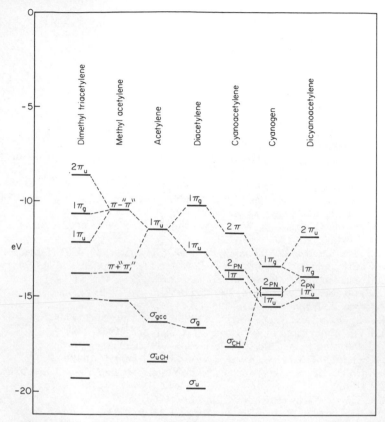

Figure 6.34 The correlation of the energy levels in acetylene, di-acetylene and related compounds produced by replacing $\equiv CH$ by $\equiv N$ (see text)

REFERENCES

1. D. W. Turner, *Advan. Phys. Org. Chem.*, **4**, 31 (1966).
2. A. Streitwieser, *J. Am. Chem. Soc.*, **82**, 4123 (1960).
3. A. D. Baker, C. Baker, C. R. Brundle and D. W. Turner, *Int. J. Mass Spec. Ion Phys.*, **1**, 285 (1968).
4. I. M. Mills, *Mol. Phys.*, **1**, 99 (1958).
5. R. K. Nesbet, *J. Chem. Phys.*, **32**, 1114 (1960).
6. E. D. Albasing and I. R. A. Cooper, *Mol. Phys.*, **4**, 353 (1961).
7. M. Krauss, *J. Chem. Phys.*, **38**, 564 (1963).
8. J. J. Sinai, *J. Chem. Phys.*, **39**, 1575 (1963).
9. R. Hoffmann, *J. Chem. Phys.*, **40**, 2047 (1963).
10. R. Moccia, *J. Chem. Phys.*, **40**, 2164 (1964).
11. W. E. Palke and W. N. Lipscomb, *J. Am. Chem. Soc.*, **88**, 2384 (1966).

12. W. C. Price, unpublished results.
13. K. Hamrin, G. Johansson, U. Gelius, A. Fahlman, C. Nordling and K. Siegbahn, *Chem. Phys. Lett.*, **1**, 613 (1968).
14. W. C. Price, *Phys. Rev.*, **47**, 444 (1935).
15. B. A. Lombos, P. Sauvageau and C. Sandorfy, in the press.
16. E. N. Lassettre, A. Skerbele and M. A. Dillon, *J. Chem. Phys.*, **46**, 4536 (1967).
17. W. A. Chupka and J. Berkowitz, *J. Chem. Phys.*, **47**, 2921 (1967).
18. M. I. Al-Joboury, *Ph.D. Thesis*, London University, 1964.
19. H. Berthod, *Compt. Rend.*, **249**, 1354 (1959).
20. U. Kaldor and I. Shavitt, *J. Chem. Phys.*, **48**, 191 (1968).
21. J. W. Moskowitz, *J. Chem. Phys.*, **43**, 60 (1965); **45**, 2338 (1966).
22. J. W. Schulmann, J. W. Moskowitz and C. Hollster, *J. Chem. Phys.*, **46**, 2759 (1967).
23. J. L. Whitten, *J. Chem. Phys.*, **44**, 359 (1966).
24. M. B. Robin, R. R. Hart and N. A. Kuebler, *J. Chem. Phys.*, **44**, 1803 (1966).
25. R. S. Mulliken, *J. Chem. Phys.*, **7**, 121 (1939).
26. J. G. Aston, G. Szasz, H. W. Woolley and F. G. Brickwedde, *J. Chem. Phys.*, **14**, 67 (1946).
27. W. B. Smith and J. L. Massingill, *J. Amer. Chem. Soc.*, **83**, 4301 (1961).
28. W. C. Price and A. D. Walsh, *Proc. Roy. Soc.* (*London*), *Ser. A*, **174**, 220 (1940).
29. T. M. Sugden and A. D. Walsh, *Trans. Faraday Soc.*, **41**, 76 (1945).
30. K. Watanabe, *J. Chem. Phys.*, **26**, 542 (1957).
31. B. Brehm, *Z. Naturforsch*, **21a**, 196 (1966).
32. M. J. S. Dewar and S. D. Worley, *J. Chem. Phys.*, **49**, 2454 (1968).
33. M. I. Al-Joboury and D. W. Turner, *J. Chem. Soc.*, **1964**, 4434.
34. A. C. Parr and F. A. Elder, *J. Chem. Phys.*, **49**, 2659 (1968).
35. N. C. Baird and M. J. S. Dewar, *Theoret. Chim. Acta*, **9**, 1 (1967).
36. H. Kato, H. Konishi, H. Yamabe and T. Yonezawa, *Bull. Chem. Soc. Japan*, **40**, 2761 (1967).
37. R. K. Harris, *Spectrochim. Acta*, **20**, 1129 (1964).
38. D. A. Hutchinson, *Trans. Faraday Soc.*, **59**, 1695 (1963).
39. K. B. Wiberg and B. J. Nist, *J. Am. Chem. Soc.*, **83**, 1226 (1961).
40. L. H. Sutcliffe and A. D. Walsh, *J. Chem. Soc.*, **1952**, 899.
41. V. H. Dibeler, R. M. Reese and J. L. Franklin, *J. Amer. Chem. Soc.*, **83**, 1813 (1961).
42. J. E. Collin and J. Delwiche, *Can. J. Chem.*, **45**, 1883 (1967).
43. C. Baker and D. W. Turner, *Proc. Roy. Soc.* (*London*), *Ser. A*, **308**, 19 (1968).

CHAPTER 7

Three-Membered Ring Compounds

CYCLOPROPANE AND ITS DERIVATIVES, ETHYLENE OXIDE

(A) Cyclopropane

The electronic structure of cyclopropane has been discussed many times in the past (see for example References 1–7) but the photoelectron spectrum can be interpreted using Walsh's[5] and Sugden's[4] descriptions of the molecular orbitals as a starting point. A comparison of the spectrum has also been made with the optical spectra and with Gaussian orbital SCF calculations.[8] These calculations predict four orbitals with energies above $-21 \cdot 2$ eV, and lead to an electronic structure for cyclopropane which can be written as

$$KKK \, (2a_1')^2 \, (2e')^4 \, (1a_2'')^2 \, (3a_1')^2 \, (1e'')^4 \, (3e')^4 \dots {}^1A_1'$$

This however involves reversing the order of the first and second, and the third and fourth orbitals of the configuration suggested by Sugden.[4] A schematic representation of the shapes of these orbitals is given in Figure 7.1. Loss of an electron from the doubly degenerate pairs of orbitals $(3e')$ and $(1e'')$ should produce ions in states which are unstable with respect to Jahn–Teller distortion (cf. methane and allene (Chapter 6) and the methyl halides (Chapter 8)). The double maximum associated with the first band in the spectrum (10–12 eV, Figure 7.2) may be a result of this distortion. The second band (12–14 eV) should also show this effect, but it is not so obvious from the spectrum though there is some evidence of an asymmetrical broadening.

Associated with the first band is a progression of peaks corresponding to a vibrational frequency of 480 cm^{-1}, probably attributable to the ring deformation v_{11} which has a frequency of 868 cm^{-1} in the molecular ground state.[9] Deuteration failed to provide information on this point as no fine structure could be resolved in the first band of the cyclopropane-d_6 spectrum. A vibrational frequency of 490 cm^{-1} has been observed in two of the Rydberg levels of the cyclopropane molecule.[8]

The fine structure associated with the fourth band (adiabatic I.P. 16·5 eV) indicates a vibrational frequency of 1130 cm^{-1}, which reduces to 900 cm^{-1} on deuteration. This is consistent with the excitation of a CH_2 mode, expected as a result of electron loss from a π type (H—C—H bonding) orbital.

(B) Ethylene oxide

This spectrum (Figures 7.3–7.6) bears superficial resemblance in most of the bands to that of cyclopropane. Indeed, the 16·5 eV band of ethylene oxide correlates with that at 15·7 eV in cyclopropane, and also the 17·5 eV ethylene oxide band with the 16·5 eV cyclopropane band. However, MO calculations[8] predict a scrambling of the upper σ and π orbital levels of the hydrocarbon on the replacement of —CH_2— by —O— (Figure 7.7). Undoubtedly the first band ($2b_2\pi$, 10·57 eV) is produced by electrons from the orbital derived from one of the cyclopropane ($1e''$) orbitals, but now largely localized on the oxygen atom. Two vibrational modes are excited in this state of the ion, with frequencies of 806 and 1130 cm^{-1}. The first of these can be assigned to the symmetric ring-deformation v_5 (877 cm^{-1} in the molecule).[9] The second is likely to be the symmetric CH_2 'scissor' mode v_2 (1490 cm^{-1} in the molecule),[9] but the ring breathing mode v_3 (1120 cm^{-1} in the molecule)[9] cannot be ruled out.

(C) Ethylenimine

The spectrum of this molecule (Figure 7.8) is somewhat different from the two preceding isoelectronic compounds, but the orbitals can be related to the others in the series (Figure 7.7).

Some of the Rydberg states[8] of the molecule exhibit a vibration of frequency 709 cm^{-1}, which is observed in the first band of the spectrum, at ~ 700 cm^{-1}. Once again this is probably the symmetric ring deformation, which has a frequency of 856 cm^{-1} in the molecular ground-state.[10]

(D) Derivatives of Cyclopropane

As in the case of benzene derivatives (see Chapter 11), the introduction of a substituent in place of one of the hydrogen atoms of cyclopropane introduces the possibility of interaction of the ring orbitals with those of the substituent. However, unlike benzene, the highest occupied orbital is σ, the e' 'external σ orbital'. Reference to Figure 7.1 shows that np atomic substituent orbitals may overlap so as to affect the energy of one of the degenerate (e') orbitals more than the other, and so the degeneracy can be removed.

The spectra of three cyclopropane derivatives, viz. the bromide, amine, and cyanide, are shown in Figures 7.9–7.12, and their ionization energies compared in Figure 7.7.

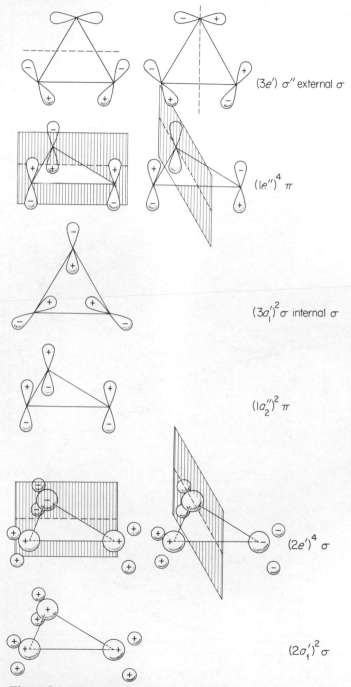

$(3e')$ σ'' external σ

$(1e'')^4$ π

$(3a_1')^2$ σ internal σ

$(1a_2'')^2$ π

$(2e')^4$ σ

$(2a_1')^2$ σ

Figure 7.1   The occupied orbitals of cyclopropane illustrated
in terms of their constituent atomic orbitals

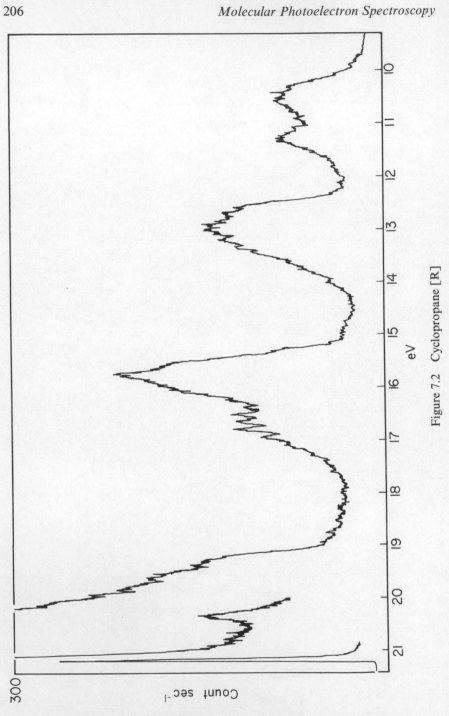

Figure 7.2   Cyclopropane [R]

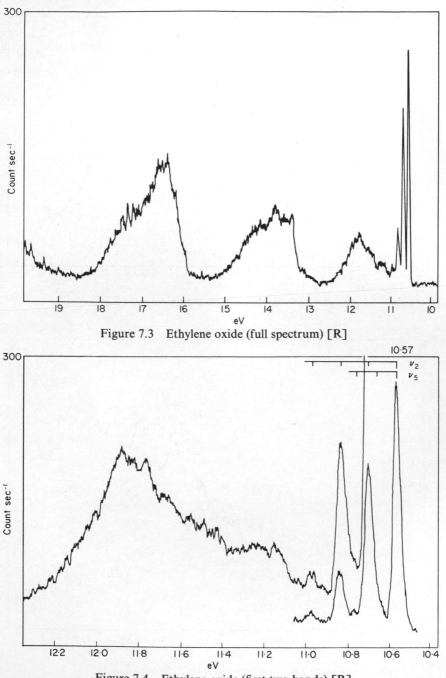

Figure 7.3   Ethylene oxide (full spectrum) [R]

Figure 7.4   Ethylene oxide (first two bands) [R]

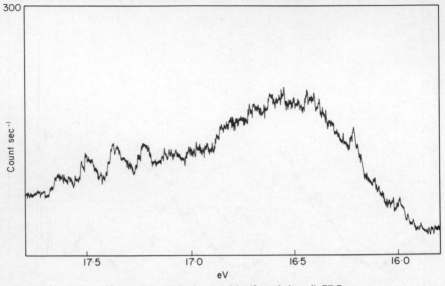

Figure 7.5   Ethylene oxide (fourth band) [R]

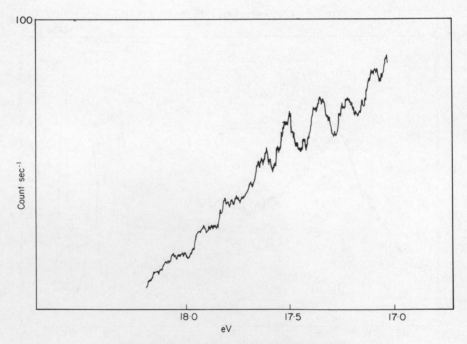

Figure 7.6   Ethylene oxide (fifth band) [R]

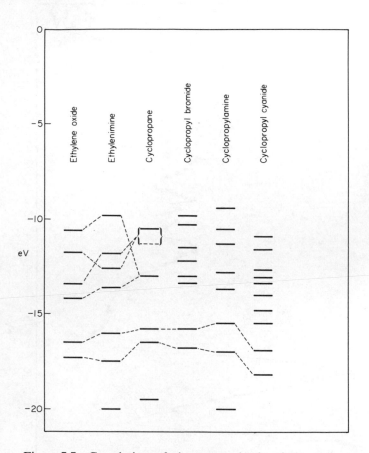

Figure 7.7   Correlation of the energy levels of the cyclopropane and related molecules

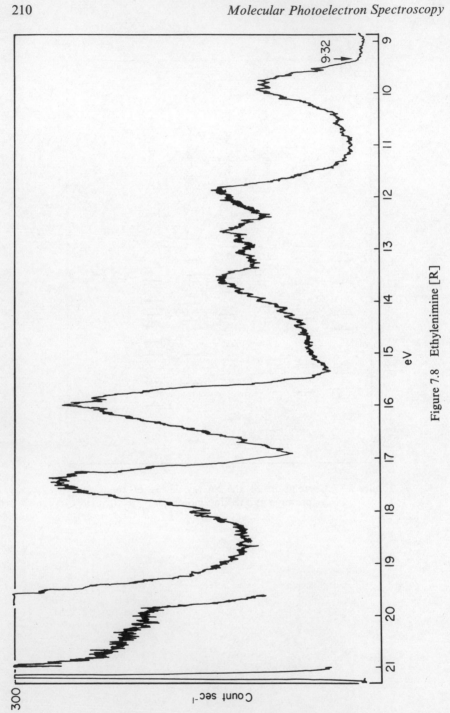

Figure 7.8 Ethylenimine [R]

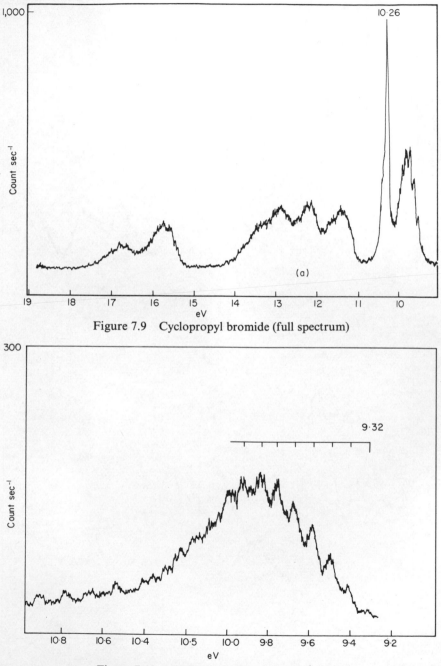

Figure 7.9   Cyclopropyl bromide (full spectrum)

Figure 7.10   Cyclopropyl bromide (first band)

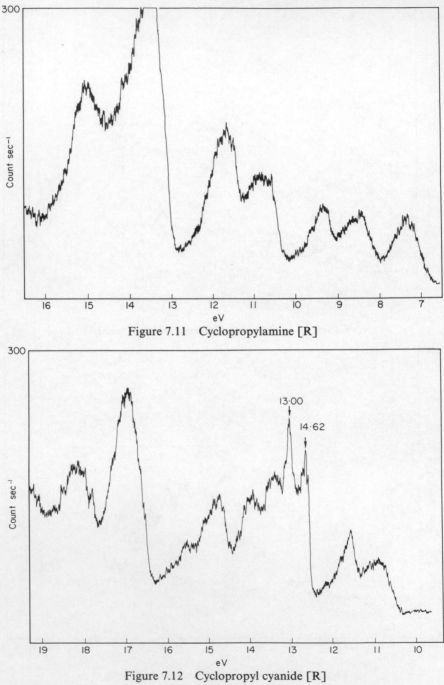

Figure 7.11   Cyclopropylamine [R]

Figure 7.12   Cyclopropyl cyanide [R]

REFERENCES

1. A. D. Walsh, *Nature*, **159**, 165 (1947).
2. R. Robinson, *Nature*, **159**, 400 (1947).
3. C. A. Coulson and W. E. Moffitt, *Phil. Mag.*, **40**, 1 (1949).
4. T. M. Sugden, *Nature*, **160**, 367 (1947).
5. A. D. Walsh, *Trans. Faraday Soc.*, **45**, 179 (1949).
6. R. D. Brown and V. G. Krishna, *J. Chem. Phys.*, **45**, 1482 (1966).
7. J. W. Raymonda and W. T. Simpson, *J. Chem. Phys.*, **47**, 430 (1967).
8. H. Basch, M. B. Robin, N. A. Kuebler, C. Baker and D. W. Turner, Submitted to *J. Chem. Phys.*
9. G. Herzberg, *Molecular Spectra and Molecular Structure*, Vol. III, Van Nostrand, New York, 1966.
10. W. J. Potts, *Spectrochim. Acta*, **21**, 511 (1965).

CHAPTER 8

Alkyl and Alkenyl Halides

Photoelectron spectra of compounds containing halogen atoms often show one or more fairly sharp bands, with more than 90% of the total transition probability in the $0 \leftarrow 0$ component. Such bands result from the removal of one of the halogen 'lone-pair' electrons, which in the molecular ground state are in an orbital which is almost atomic in character. In simple saturated monohalides, these orbitals are almost nonbonding, but in the case of polyhalides, varying bonding properties are observed for the halogen orbitals.

The spectra of fluorine containing compounds often lack these sharp bands, apparently because fluorine $2p$ orbitals are able to participate more fully in the formation of p-π bonds. The large electronegativity of the fluorine atom has such a considerable effect on the electronic structures of fluorine-containing molecules that their spectra differ markedly from those of related molecules containing the other halogens. For this reason, the spectra of the fluoromethanes will be discussed separately.

1. SATURATED COMPOUNDS

(A) Methyl Chloride, Bromide and Iodide

The electronic structure of the ground state of the methyl halides can be derived from that of methane, which can be written as

$$(1s_C)^2 \, (sa_1)^2 \, (pt_2)^6 : {}^1A_1$$

Replacement of one of the hydrogens by a halogen reduces the T_d symmetry to C_{3v}, and the (pt_2) orbital splits into $(\pi_e)^4 \, (\sigma a_1)^2$, the (σa_1) orbital being the C–halogen bond. Inclusion of the outer orbitals of the halogen gives the electronic structure[1]

$$(1s_C)^2 \, (sa_1)^2 \, (ns_X a_1)^2 \, (\pi_e)^4 \, (\sigma a_1)^2 \, (np\pi_{eX})^4 : {}^1A_1$$

where $n = 3$, 4 or 5 for X = Cl, Br or I.

The highest occupied orbital is largely localized on the halogen atom, though there may be some X—H antibonding and C—H bonding character.[2] Removal of an electron from this orbital gives an ion in the ^2E state which is doubly degenerate, and should be susceptible to both spin–orbit and Jahn–Teller effects. For the interpretation of such ionic systems, which have an odd number of electrons, it is necessary to use the double point group constructed from the C_{3v} group and the electron spin $^2\Gamma$. This new group C_{3v}^* ($= C_{3v} \times ^2\Gamma$), a so-called 'spinor group', has two additional representations E_1 and $E_{3/2}$, and twice as many elements as C_{3v}.[3,4] In molecules where spin–orbit coupling is strong, (a) a ^2D state splits into $^2E_{1/2}$ and $^2E_{3/2}$, which cannot be split further by Jahn–Teller effects since each is nondegenerate— a Kramers doublet. In molecules where spin–orbit and Jahn–Teller effects are of comparable magnitudes, (b) a displacement of the ^2E energy levels from the symmetric conformation occurs and two new levels form, but whereas in simple Jahn–Teller distortions already discussed, the potential energy curves for the two states touch, they are now pushed apart by the spin–orbit interaction. The lower of the two states has two off-axis minima while the upper state has one minimum at the symmetric conformation.

In the methyl halides, Mulliken[1] has predicted the energy separation between the $^2E_{3/2}$ and $^2E_{1/2}$ states to be 0·08 eV in CH_3Cl^+, 0·32 eV in CH_3Br^+ and 0·625 eV in CH_3I^+ and verification has been forthcoming from the absorption spectra.[5]

Turning now to the photoelectron spectra of the methyl halides (Figures 8.3, 8.5 and 8.7), if we take the first bands throughout to be ^2E, from their appearance it seems that CH_3Br and CH_3I are instances of case (a) above, where the $^2E_{3/2}$ and $^2E_{1/2}$ states are well separated, and in fact the magnitudes of the spin–orbit splittings agree with the predictions. The first band in the CH_3Cl spectrum (Figure 8.4) is complex, particularly the early members. This is consistent with a case (b) situation, since the double minimum of the $^2E_{3/2}$ state would result in distortions of the vibrational energy levels.

Table 8.1 lists the spin–orbit splittings which have been observed by different techniques, together with Mulliken's predictions.

The absorption spectra[5] of the methyl halides indicate excitation of the CH_3 deformation (ν_2), each transition being accompanied by one quantum of ν_3, the C–halogen stretching mode. A similar interpretation is possible for the $^2E_{1/2}$ state of CH_3Br^+ (Figure 8.6, Table 8.1), but in the $^2E_{3/2}$ state, in addition to ν_2, a frequency of 810 cm^{-1} appears, which is considerably above the molecular frequency of ν_3. If it does refer to ν_3, then the difference between the ionic and molecular frequencies is surprising. In the spectrum of CH_3I, a frequency of 1290 cm^{-1} is observed in both ^2E states, this value being slightly greater than the molecular frequency of ν_2, in accordance with the slight I—H antibonding character of the orbital.

Table 8.1   Ionization potentials and separations between the $^2E_{1/2}$ and $^2E_{3/2}$ states of the CH_3X^+ ions, where X = Cl, Br or I

Reference and technique	State of ion	CH_3Cl^+		CH_3Br^+		CH_3I^+	
		I.P.	Separation	I.P.	Separation	I.P.	Separation
1 (Prediction)	$^2E_{3/2}$		0·08		0·32		0·625
	$^2E_{1/2}$						
5 (S.)	$^2E_{3/2}$	11·22	0·08	10·540	0·315	9·537	0·628
	$^2E_{1/2}$	11·30		10·855		10·165	
6 (P.E.S.)	$^2E_{3/2}$	11·28		10·54	0·31	9·55	0·61
	$^2E_{1/2}$			10·85		10·16	
7 (P.E.S.)	$^2E_{3/2}$	11·29	0	10·55	0·31	9·56	0·63
	$^2E_{1/2}$	11·29		10·86		10·19	
8 (E.I.)	$^2E_{3/2}$	11·42	0	10·53	0·32	9·51	0·58
	$^2E_{1/2}$	11·42		10·85		10·09	
9 (E.I.)	$^2E_{3/2}$	11·3	0·1	10·5	0·3	9·5	0·5
	$^2E_{1/2}$	11·4		10·8		10·0	
10 (P.I.)	$^2E_{3/2}$	11·265	0·075	10·528	0·329	9·550	0·57
	$^2E_{1/2}$	11·340		10·857		10·12	
11 (P.I.)	$^2E_{3/2}$					9·55	0·60
	$^2E_{1/2}$					10·15	

S. = Spectroscopic   P.E.S. = Photoelectron Spectroscopy.
E.I. = Electron Impact.   P.I. = Photoionization.

As already mentioned, the vibrational levels of the $^2E_{3/2}$ state of CH_3Cl^+ are likely to be distorted, but it is possible that the levels of the $^2E_{1/2}$ state, particularly the higher ones, are not distorted to such an extent. It is perhaps significant therefore, that the four vibrational peaks at highest I.P. in Figure 8.6 can be analysed in terms of frequencies of 1170 cm^{-1} (v_2', $v_2'' = 1355$ cm^{-1} and 520 cm^{-1} (v_3', $v_3'' = 733$ cm^{-1}).

The frequencies excited in the ^2E ionic states of methyl bromide and iodide are collected in Table 8.2.

The band in the spectra which corresponds to electrons from the (σa_1) orbital is the first broad band in each case (i.e. at 14·4, 13·5 and 12·5 eV respectively). Hartree–Fock calculations[12] give the energy of the (σa_1) orbital in CH_3Cl as 14·4 eV. The energy required to remove an electron from this orbital is directly related to the energy of the C–halogen band, and if the vertical I.P.'s of the (σa_1) orbitals are plotted against corresponding bond energies,[13] a linear relationship is obtained for these compounds (Figure 8.8).

The remaining broad band in the spectra corresponds to electron loss from the (π_e) orbital, which is largely localized on the methyl group, though there is some C–halogen bonding. This orbital level is doubly degenerate, and the removal of an electron should produce an ion which is unstable, and should undergo Jahn–Teller distortion. The (π_e) bands in the spectra are broad, and show a slight shoulder on the high I.P. side. This may be the result of the expected distortion, the degeneracy of the orbital being lifted in the ion.

The I.P.'s of methyl chloride, bromide and iodide are correlated in Figure 8.27.

Table 8.2   Vibrational frequencies excited in CH_3Br^+ and CH_3I^+

	State	Frequency in ion (cm^{-1})	Frequency in mol (Ref. 3)	Assignment	Description of mode
CH_3Br^+	$^2\dot{E}_{3/2}$	1210	1305	$v_2(a_1)$	CH_3 deformation
$(\tilde{X})$		810			
	$^2E_{1/2}$	1290	1305	$v_2(a_1)$	CH_3 deformation
		560	611	$v_3(a_1)$	C—Br stretching
CH_3I^+	$^2E_{3/2}$	1290	1251	$v_2(a_1)$	CH_3 deformation
$(\tilde{X})$	$^2E_{1/2}$	1290	1251	$v_2(u_1)$	CH_3 deformation

(B) Methylene Dichloride, Dibromide and Diiodide

The electronic structure of these dihalo-methanes can be derived from that of the methyl halides by considering the effect of a reduction in symmetry from C_{3v} (CH_3X) to C_{2v} (CH_2X_2).

The degenerate (π_e) orbital of the C_{3v} system splits into b_1 and b_2 orbitals in the C_{2v} case. The b_1 orbital will be C—H bonding and the b_2 orbital, C—X bonding. The halogen p orbitals in the dihalo-molecules can interact and form four new molecular orbitals (a_1), (a_2), (b_1) and (b_2).

Of these, only (a_2) is strictly nonbonding, since the others are of correct symmetry to overlap with the carbon $2p$ orbitals.

Thus the electronic structure of CH_2X_2 can be written as

$$\ldots (a_1)^2 \ (b_1)^2 \ (a_1)^2 \ (b_2)^2 \ [(a_1)^2 \ (a_2)^2 \ (b_1)^2 \ (b_2)^2] : {}^1A_1$$

$$2s_C \ (2p_C, np\sigma_X, 1s_H) \qquad np\pi_X$$

It is not intended to specify the ordering of the orbitals within the squared bracket, however.

From the photoelectron spectra (Figures 8.13, 8.15, 8.16) it can be seen that the four 'halogen' orbitals predicted by group theory are in fact detected in CH_2I_2 (Figure 8.16) and CH_2Br_2 (Figure 8.15). Since the (a_2) orbital is the least bonding, it is likely to be associated with the sharpest, most intense bands in the spectra, and therefore can probably be assigned to the third or fourth bands. In the spectrum of CH_2Cl_2, however, only two bands appear in the corresponding region (11–13 eV). The areas under these two bands are approximately equal, and so it may be that each is the result of two overlapping bands, the chlorine nonbonding orbitals forming two accidentally degenerate pairs. The nonbonding (a_2) orbital can possibly be assigned to the sharp peak at 12·18 eV.

The first band in the CH_2Cl_2 spectrum (Figure 8.14), at adiabatic I.P. 11.31 eV, contains a short vibrational series with $\Delta G = 650$ cm^{-1}. This is probably associated with the symmetric C—Cl stretching mode v_3' (where $v_3'' = 713$ cm^{-1}). A frequency of 670 cm^{-1} has been obtained for this mode from the absorption spectrum.[14]

The spectral bands corresponding to electrons ejected from the bonding (a_1), (b_1) and (b_2) orbitals can clearly be seen in the 12·5–17 eV region of the spectra. A rigorous assignment of each band is not possible from the spectrum, but (b_2), which is CX_2 bonding, might be expected to have the lowest I.P., and (b_1), which is CH_2 bonding, the highest I.P. To support this latter assignment it can be seen from the spectra that the orbital with the highest I.P. of these three undergoes the smallest changes in I.P. through the series CH_2Cl_2, CH_2Br_2, CH_2I_2, and is probably therefore not directly involved in C—X bonding. Both (a_1) and (b_2) are partly C—X bonding, and show the expected decrease in I.P. through the series.

The weak band in the 19–20 eV region is probably due to electron loss from the (a_1) orbital which may approximate to $2s_C$.

(C) Chloroform, Bromoform and Iodoform

As these molecules are of C_{3v} symmetry, the bonding orbitals will be similar to those of the methyl halides, namely (σa_1) and (π_e). In these cases however, the (σa_1) orbital will be C—H bonding, and the (π_e) orbital C—X bonding. Consequently, the I.P. of the (π_e) orbital might be expected to be lower than that of (σa_1). This is supported by the areas under the relevant spectral bands in the 14–18 eV region of Figures 8.20 and 8.21 that at lower I.P., $(\pi_e)^4$, having approximately twice the area of the other, $(\sigma a_1)^2$.

The halogen $p\pi$ orbitals can, in the molecule, be combined to give four filled molecular orbitals a_1, a_2, e and e in the spectrum of chloroform (Figure 8.20), only three bands appear in the expected region (11–14 eV), therefore two of the orbitals are apparently accidentally degenerate. The areas of the peaks are in the ratio of approximately 2:3:1, and this, and also comparison with the $CHBr_3$ spectrum (see below), suggests that the 11·5 eV band is of (a) symmetry, the 12 eV band is in fact overlapping (a) and (e) bands, and the 12·8 eV band is (e). It is not possible at this stage to obtain from the spectrum the relative order of (a_1) and (a_2). (a_2) is, however, completely nonbonding, and comparison with methylene chloride suggests that its I.P. might be higher than that of (a_1).

An adequate explanation of the chloroform spectrum is thus obtained in terms of the molecular orbitals, (a_1), (a_2) and (e), which are irreducible representations of the C_{3v} point group. These refer only to even electron systems, i.e. systems without spin. For interpretation of the photoelectron spectra of bromoform and iodoform, where spin–orbit splitting is significant, it is again necessary to use the double group. In the CHX_3^+ ion, the states $^2A_1$, $^2A_2$ and ^2E become $E_{1/2}$, $E_{1/2}$ and $E_{1/2} + E_{3/2}$, respectively. In $CHCl_3^+$, it appears that the $E_{1/2}$ and $E_{3/2}$ states are approximately degenerate as seen. However, in $CHBr_3^+$, this is no longer the case, and a total of six peaks result, the $E_{1/2}$ and $E_{3/2}$ states having moved apart. The sharp intense peak at I.P. 11·25 eV in the $CHBr_3$ spectrum, and 10·29 eV in that of CHI_3 results from electron loss from the most nonbonding of the orbitals, which is therefore probably related to the $^2A_2$ (or $E_{1/2}$) state of $CHCl_3$ (see above).

(D) Carbon Tetrachloride and Tetrabromide

In the spectra of these two molecules (Figures 8.25 and 8.26), the band in the 14–17 eV region can clearly be assigned to the $(pt_2)^6$ orbital, which is C—X bonding. The ion formed by the removal of an electron from this orbital should be susceptible to Jahn–Teller distortion (cf. methane), but this is not apparent from the spectrum, since the effect will be small because of the large mass of Cl and Br atoms.

The band at about 20 eV in the spectra may be due to electrons from the (a_1) orbital which is approximately $2s_C$.

The halogen $p\pi$ orbitals can be combined to give three occupied molecular orbitals which, in the molecule, correspond to the representations (e), (t_1) and (t_2), both (t) orbitals being triply degenerate and (e) doubly degenerate. Turning once more to the spinor group T_d^*, the $^2(E_1)$ $^2(T_1)$ and $^2(T_2)$ states transform to $G_{3/2}$, $E_{1/2} + G_{3/2}$ and $E_{5/2} + G_{3/2}$ respectively, of which the G states are doubly degenerate. The three main bands in the CCl_4 spectrum clearly are associated with ^2E, $^2T_1$ and $^2T_2$ states, the one at 13·5 eV being ^2E, though $^2T_1$ and $^2T_2$ cannot be distinguished at this stage. The difference between the energies of the related E and G orbitals is dependent upon the magnitude of the spin–orbit splitting, and since this is small for Cl, the individual bands are not resolved. However, the G states, being doubly degenerate, are susceptible to Jahn–Teller distortion, and so the breadth of the main peaks, and the small peak on the high I.P. side of the 12·6 eV peak, may be due to unresolved, or partially resolved, Jahn–Teller components of the G states.

On going to CBr_4, the larger spin–orbit splitting brings about separation of the E and G states, while the much smaller Jahn–Teller splitting of the G states can also be seen.

The situation regarding spin–orbit splitting in a ^2T state is similar to that in a ^2P state, as indicated schematically below,

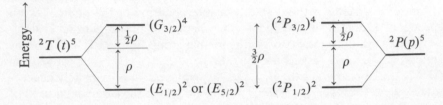

the separation being $\frac{3}{2}\rho$.

The separation (ρ_1) between the states derived from $^2T_1$, viz. $E_{1/2}$ and $G_{3/2}$, is expected[15] to be about $\frac{1}{2}\rho_p$, where ρ_p is the coupling constant in a bromine atom, and the $E_{5/2} - G_{3/2}$ separation (ρ_2) should be ρ_p.

Therefore the splittings expected are $\frac{3}{4}\rho_p$ in the $^2T_1$ state, and $\frac{3}{2}\rho_p$ in the $^2T_2$ state. Since ρ_p is 0·305 eV, splittings of 0·23 and 0·46 eV should be observed. From the CBr_4 spectrum the splitting associated with the higher ^2T state is 0·32 eV, and with the lower ^2T state, 0·54 eV. Although the absolute magnitudes are not in agreement, it seems probable therefore that the higher ^2T state is $^2T_1$, and the lower state, $^2T_2$.

The changes in orbitals, on successive halogen substitution, which have been discussed in related groups of molecules, can clearly be seen by considering the series CH_3Br, CH_2Br_2, $CHBr_3$, CBr_4, as illustrated as an energy level diagram (Figure 8.28).

(E) Fluoromethanes

The spectra of the compounds formed by substitution of one or more of the hydrogen atoms of methane by a fluorine atom must be discussed separately as they differ markedly in appearance from the spectra of the corresponding chlorine, bromine and iodine substituted compounds. The substitution of one hydrogen to give methyl fluoride (Figure 8.1) has the effect of splitting the methane (pt_2) orbital, as for the other methyl halides, but the I.P. of the fluorine $2p$ orbital is expected to be higher than the (σa_1) and (π_e) orbitals.[1] Mulliken[1] suggested that the first I.P. would correspond to electron loss from the (π_e) orbital. If this were so, then the first I.P. of methyl fluoride would be expected to be close to that of methane. In fact the 12·5–14·5 eV region of the methyl fluoride spectrum seems to contain two overlapping bands (Figure 8.2), one being (π_e) and the other (σa_1). The vibrational structure associated with the first band (adiabatic I.P. 12·74 eV), has spacing corresponding to a frequency of 1070 cm^{-1}, which may be the CH$_3$ deformation v_2.

The broad band at 16–18 eV must be assigned to electrons from the fluorine 2p orbitals. The closeness of the I.P. of the (π_e) orbital to that of the (pt_2) orbital of methane, and of the fluorine $(2p\pi)$ orbital to that of the fluorine atom (17·42 eV) suggests that there is only a small interaction between (π_e) and $(2p\pi)_F$.[16]

The introduction of a second fluorine atom has the effect of splitting the (π_e) orbital of methyl fluoride into a (b_1) and a (b_2) orbital in methylene fluoride.[17] One of these must be associated with the first band in the spectrum (Figures 8.10, 8.11) as it is very similar in position, appearance, and vibrational fine structure to the first band of the methyl fluoride spectrum.

The bands in the region of 14–18 eV in the spectrum of fluoroform (Figures 8.12, 8.13) are characteristic of compounds containing a —CF$_3$ group (see CF$_3$CH$_2$I Figure 8.31 and CF$_3$CHClBr Figure 8.29).

Table 8.3 Vibrational frequencies excited in CH$_3$F, CH$_2$F$_2$ and CHF$_3$.

	State of ion of molecule		Frequencies and modes observed	
CH$_3$F	$\tilde{X}$	v_2	1468 cm^{-1}	CH$_3$ deformation
CH$_3$F$^+$	$\tilde{X}$	v_2	1070 cm^{-1}	
CH$_2$F$_2$	$\tilde{X}$	v_2	1508 cm^{-1}	CH$_2$ deformation
		v_3	1079 cm^{-1}	C—F stretching
		v_4	532 cm^{-1}	CF$_2$ deformation
CH$_2$F$_2^+$	$\tilde{X}$	v_2	1120 cm^{-1}	
	$\tilde{B}$	v_3	980 cm^{-1}	
		v_4	480 cm^{-1}	
	$\tilde{C}$	v_3	680 cm^{-1}	
CHF$_3$	$\tilde{X}$	v_2	1117 cm^{-1}	C—F stretching
		v_3	697 cm^{-1}	CF$_3$ deformation
CHF$_3^+$	$\tilde{C}$	v_2	1030 cm^{-1}	
		v_3	570 cm^{-1}	
	$\tilde{D}$	v_3	480 cm^{-1}	

Frequencies for the molecule are taken from Reference 26.

The lack of vibrational fine structure on the bands in the carbon tetra-fluoride spectrum (Figure 8.24) may be due to the instability of the CF_4^+ ion. This ion is usually not detected in mass spectrometry (cf. CCl_4^+, CBr_4^+).[18,19] It is not clear from the spectrum whether the first I.P. of carbon tetrafluoride relates to electron loss from a (pt_2) type of orbital (C—F σ bonding) or from a combination of fluorine $2p$ orbitals (nonbonding). Both possibilities have been envisaged.[20–25]

The vibrational structure associated with some of the bands in the fluoro-methane spectra is collected in Table 8.3, and tentative analyses are given for the modes excited.

(F) 1,1,1-Trifluoro, 2-Bromo, 2-Chloroethane

The 14–18 eV region of the photoelectron spectrum (Figure 8.29) of this molecule shows the bands characteristic of the —CF_3 group (compare Figure 8.31). The bands between 11 and 12 eV (Figure 8.30) can be attributed to electrons from the bromine '$4p$' orbitals, the spin–orbit splitting being 0·26 eV, somewhat less than occurs in methyl bromide. The vibrational structure associated with these two bands may be analysed in terms of short series of spacings 520 cm^{-1} (first band) and 560 cm^{-1} (second band). These may be CF_3 deformation modes.[27]

The band between 12 and 13 eV can be assigned to the chlorine '$3p$' orbitals. Again no spin–orbit splitting is seen, but there is a short vibrational series with spacing corresponding to a frequency of 1170 cm^{-1}.

2. UNSATURATED COMPOUNDS

(A) Vinyl Mono Halides

Ionization potential measurements on these molecules are correlated in Figure 8.32. Many of the results were obtained using the low resolution spectrometer of May and Turner. The first I.P.'s of the fluoride, chloride and bromide (Figures 8.33–8.35) are lower than that of ethylene, possibly owing to resonance stabilization (see below). The inner I.P.'s are all increased compared with those of ethylene indicating the importance of the inductive effect of the halogen atoms. The very strong inductive effect of fluorine is reflected in that the difference between the inner I.P.'s of ethylene and vinyl fluoride is the largest of the series.

The two halogen 'lone pair' orbitals are not degenerate in the vinyl halides, since the one whose axis is perpendicular to the plane of the molecule can interact with the C=C π system (see also Halogenobenzenes, Chapter 11). This has the effect of lowering the I.P. of the ethylenic π orbitals, and raising that of the relevant halogen p orbital, at the same time destroying some of its nonbonding character. The other p orbital remains essentially nonbonding, and is responsible for the sharp band in the spectra.

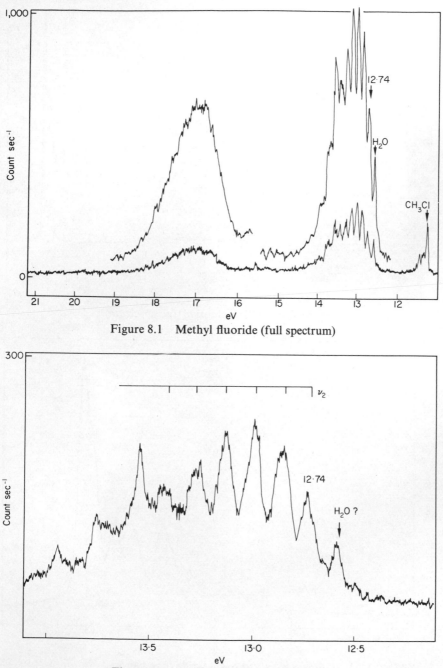

Figure 8.1   Methyl fluoride (full spectrum)

Figure 8.2   Methyl fluoride (first band)

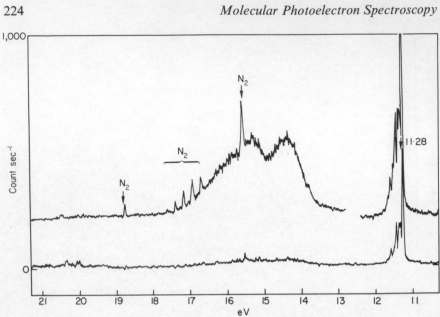

Figure 8.3   Methyl chloride (full spectrum)

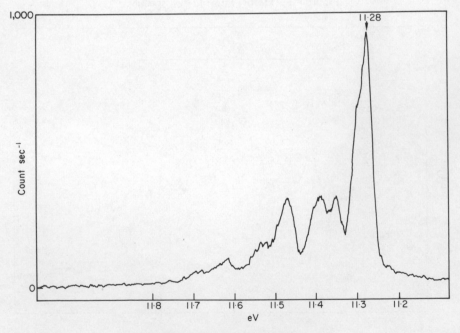

Figure 8.4   Methyl chloride (first band)

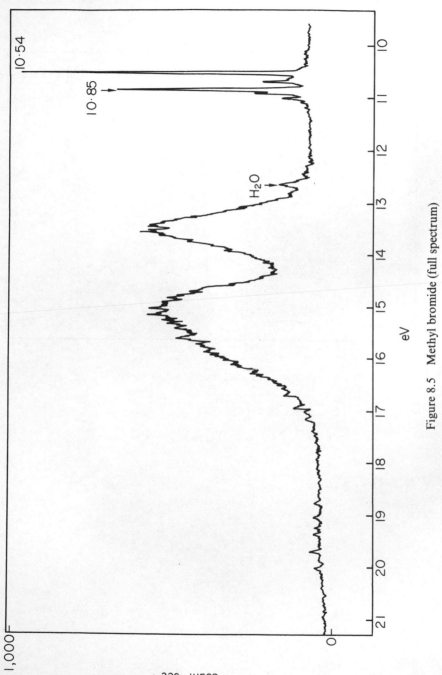

Figure 8.5   Methyl bromide (full spectrum)

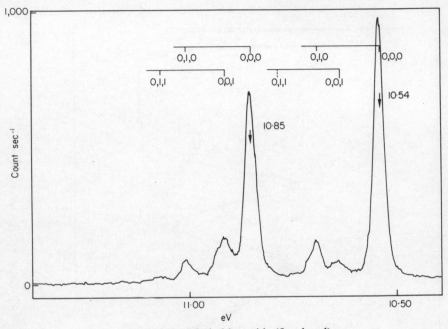

Figure 8.6   Methyl bromide (first band)

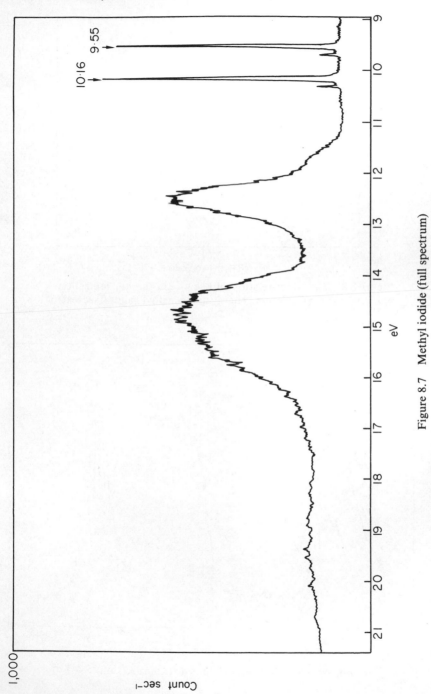

Figure 8.7 Methyl iodide (full spectrum)

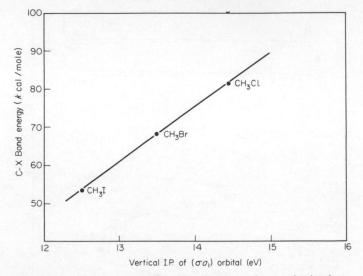

Figure 8.8   The linear relationships between the σa_1 ionization potential and the C—X bond energy in the methyl halides (see text)

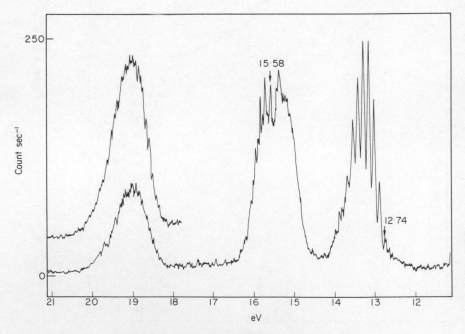

Figures 8.9   Methylene difluoride (full spectrum)

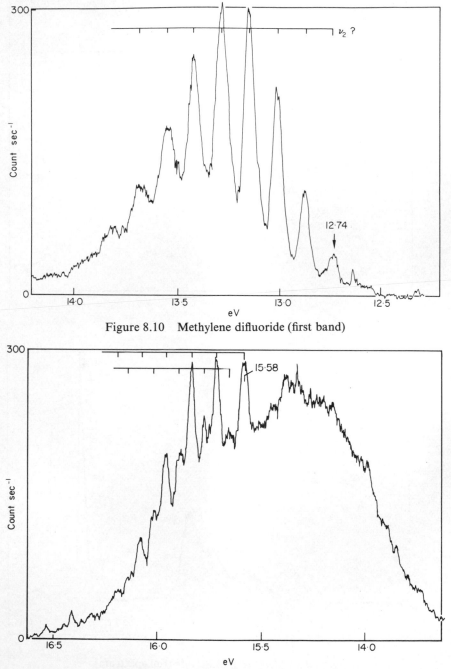

Figure 8.10   Methylene difluoride (first band)

Figure 8.11   Methylene difluoride (second band)

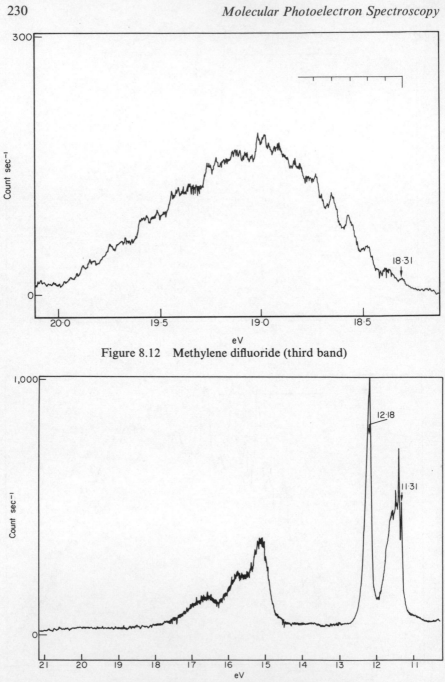

Figure 8.12   Methylene difluoride (third band)

Figure 8.13   Methylene dichloride (full spectrum)

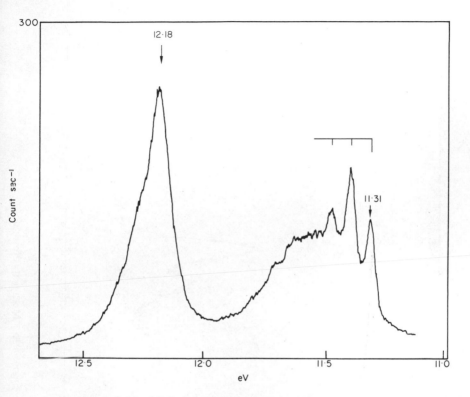

Figure 8.14   Methylene dichloride (first and second bands)

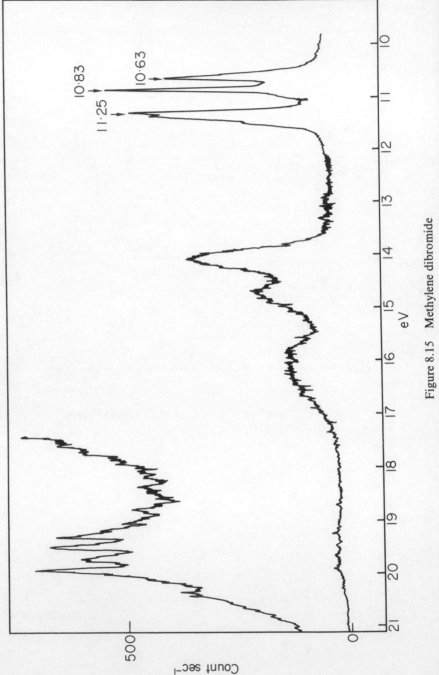

Figure 8.15 Methylene dibromide

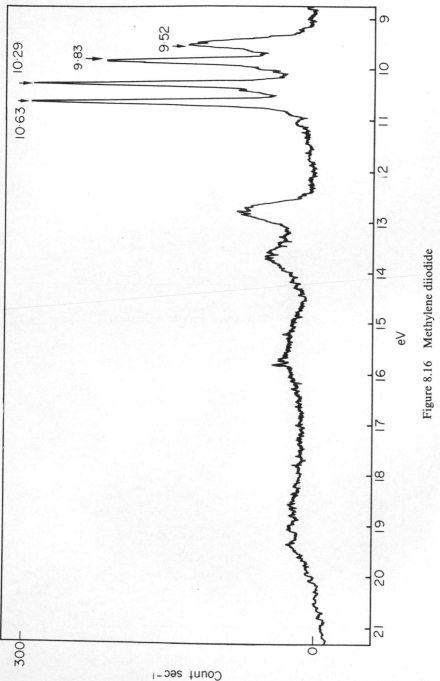

Figure 8.16 Methylene diiodide

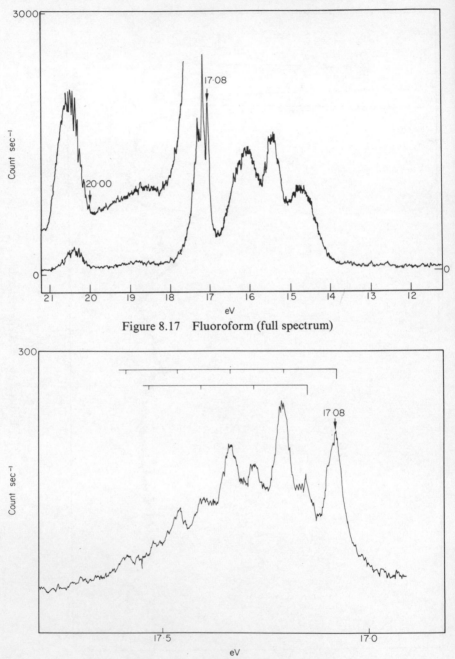

Figure 8.17   Fluoroform (full spectrum)

Figure 8.18   Fluoroform (fourth band)

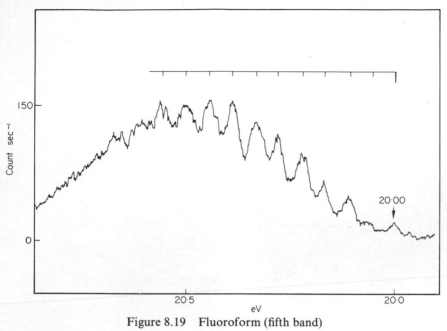

Figure 8.19　Fluoroform (fifth band)

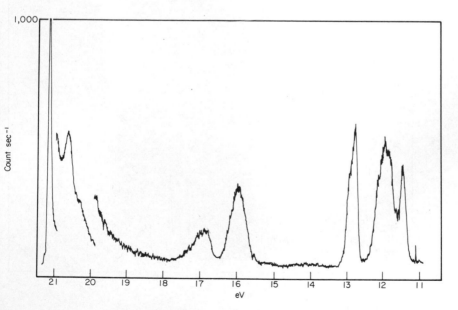

Figure 8.20　Chloroform [R]

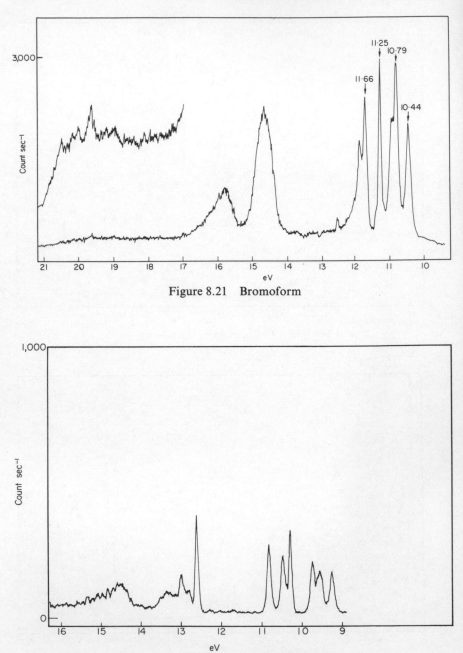

Figure 8.21   Bromoform

Figure 8.22   Iodoform (full spectrum)

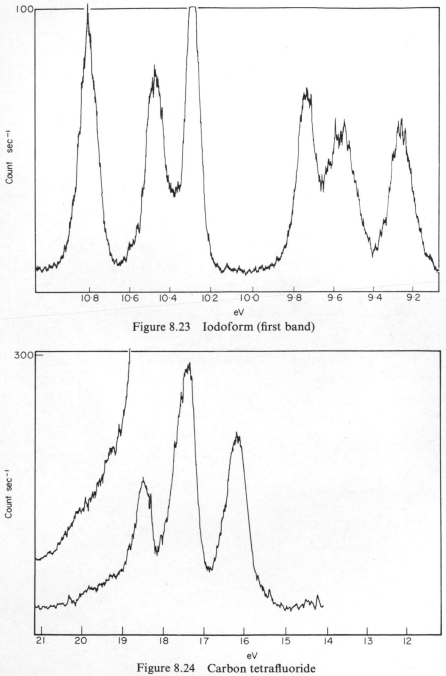

Figure 8.23   Iodoform (first band)

Figure 8.24   Carbon tetrafluoride

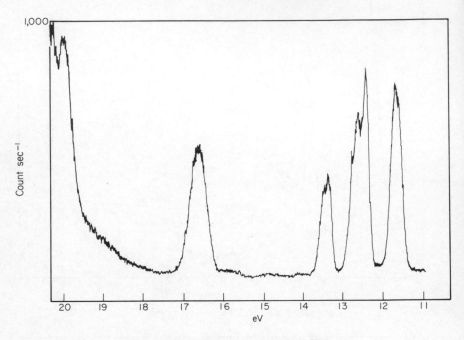

Figure 8.25   Carbon tetrachloride [R]

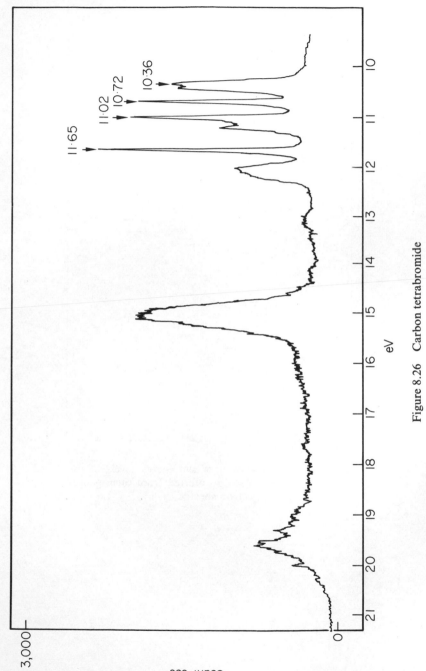

Figure 8.26 Carbon tetrabromide

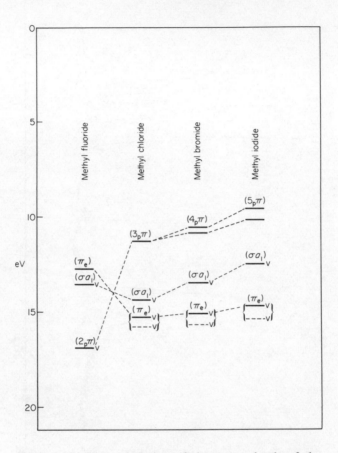

Figure 8.27   The correlation of the energy levels of the
methyl monohalide molecules inferred from their photoelectron spectra

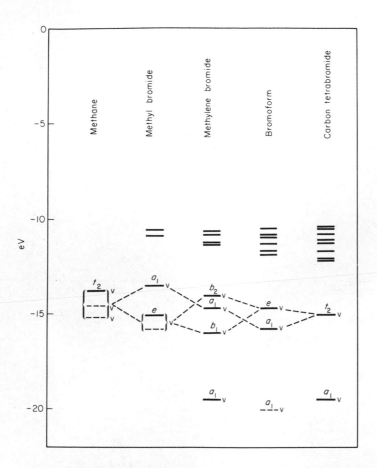

Figure 8.28 Energy-level diagram (inferred from the photo-electron spectra) showing the effect of replacing the hydrogen atoms of methane by chlorine atoms

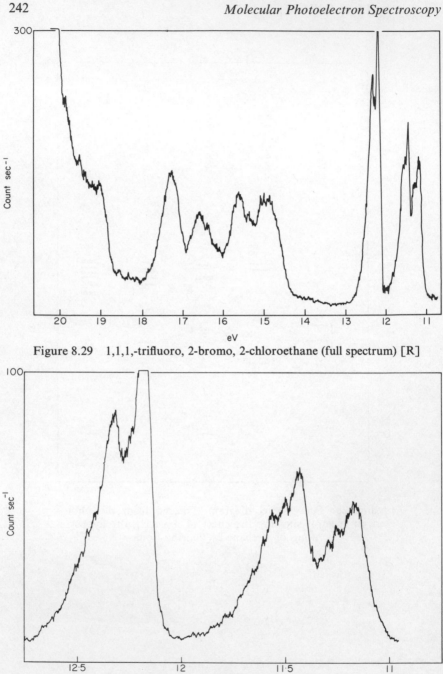

Figure 8.29   1,1,1,-trifluoro, 2-bromo, 2-chloroethane (full spectrum) [R]

Figure 8.30   1,1,1,-trifluoro, 2-bromo, 2-chloroethane (first band) [R]

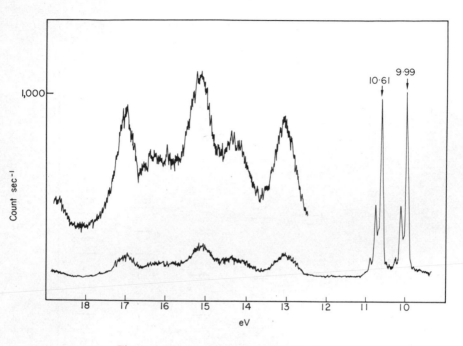

Figure 8.31   1,1,1,-trifluoro, 2-iodo ethane

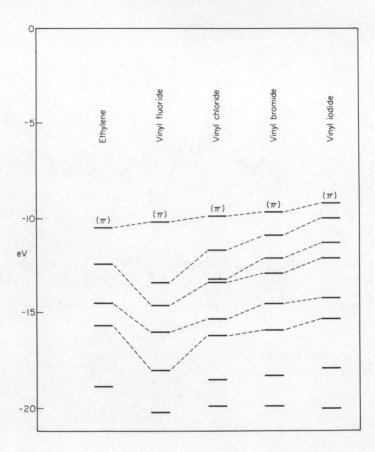

Figure 8.32   Correlation of the energy levels of the monohalogen
derivatives of ethylene

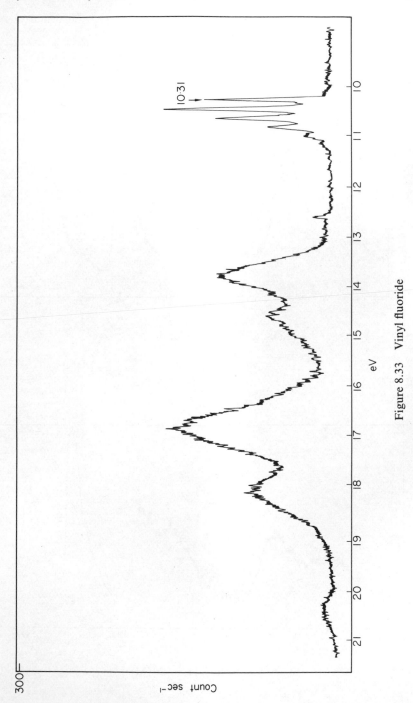

Figure 8.33 Vinyl fluoride

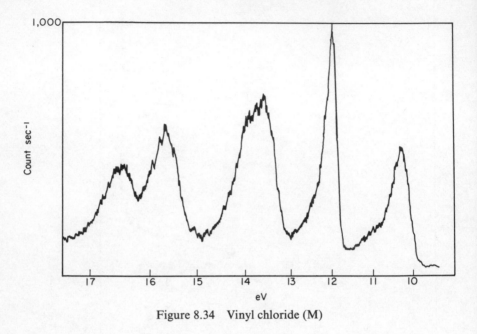

Figure 8.34   Vinyl chloride (M)

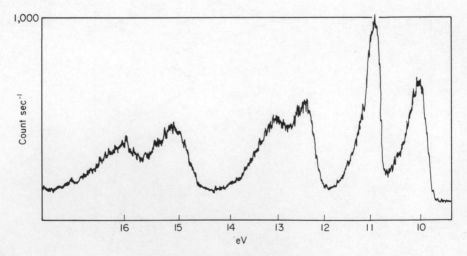

Figure 8.35   Vinyl bromide (M)

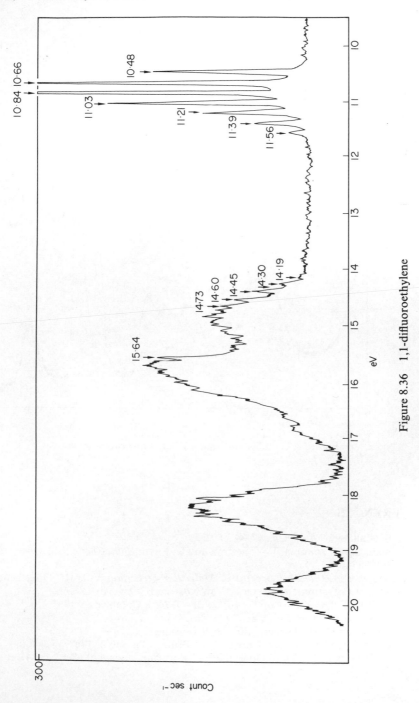

Figure 8.36 1,1-difluoroethylene

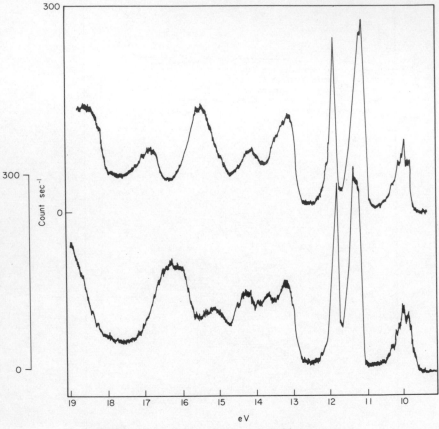

Figure 8.37   *Cis-* and *trans-* 1,3-dichloropropene

REFERENCES

1. R. S. Mulliken, *Phys. Rev.*, **47**, 413 (1935).
2. H. Kato, K. Morokuma, T. Yonezawa and K. Fukui, *Bull. Chem. Soc. Japan*, **38**, 1749 (1965).
3. G. Herzberg, *Molecular Spectra & Molecular Structure*, Vol. III, 'Electronic Spectra of Polyatomic Molecules', Van Nostrand, New York, 1966.
4. J. S. Griffith, *The Theory of Transition Metal Ions*, Cambridge, 1961.
5. W. C. Price, *J. Chem. Phys.*, **4**, 539 (1936).
6. C. Baker and D. W. Turner, Unpublished results.
7. J. S. Sandhu, *M.Sc. Thesis*, University of British Columbia, 1967.
8. D. C. Frost and C. A. McDowell, *Proc. Roy. Soc.* (*London*), Ser. *A*, **241**, 194 (1957).

9. S. Tsuda, C. E. Melton and W. H. Hamill, *J. Chem. Phys.*, **41**, 689 (1964).
10. A. J. C. Nicholson, *J. Chem. Phys.*, **43**, 1171 (1965).
11. J. D. Morrison, H. Hurzeler and M. G. Inghram, *J. Chem. Phys.*, **33**, 821 (1960).
12. M. Krauss, *Intern. Mass Spectr. Conf.*, Berlin, 1967.
13. G. Glocker, *J. Phys. Chem.*, **63**, 828 (1959).
14. C. R. Zobel and A. B. F. Duncan, *J. Am. Chem. Soc.*, **77**, 2611 (1955).
15. J. C. Green, M. L. H. Green, P. J. Joachim, A. F. Orchard and D. W. Turner, *Proc. Roy. Soc. (London), Ser. A*, In the press.
16. C. A. McDowell and B. C. Cox, *J. Chem. Phys.*, **22**, 946 (1954).
17. S. Stokes and A. B. F. Duncan, *J. Am. Chem. Soc.*, **80**, 6177 (1958).
18. V. H. Dibeler and F. L. Mohler, *J. Res. Nat. Bur. Std.*, **40**, 25 (1948).
19. R. W. Kiser and D. L. Holrock, *J. Am. Chem. Soc.*, **87**, 922 (1965).
20. G. Moe and A. B. F. Duncan, *J. Am. Chem. Soc.*, **74**, 3140 (1952).
21. G. R. Cook and B. K. Ching, *J. Chem. Phys.*, **43**, 1794 (1965).
22. C. A. Coulson and H. I. Strauss, *Proc. Roy. Soc. (London), Ser. A*, **269**, 443 (1962).
23. D. W. Davies, *Chem. Phys. Letters*, **2**, 173 (1968).
24. D. C. Frost, F. G. Herring, C. A. McDowell, M. R. Mustafa and J. S. Sandhu, *Chem. Phys. Letters*, **2**, 663 (1968).
25. P. J. Bassett and D. R. Lloyd, *Chem. Phys. Letters*, **3**, 22 (1969).
26. E. L. Pace, *J. Chem. Phys.*, **18**, 881 (1950).
27. R. Theimer and J. R. Nielsen, *J. Chem. Phys.*, **27**, 887 (1957).

CHAPTER 9

Acrolein and Some of Its Derivatives

1. ACROLEIN

The ultraviolet absorption spectrum of acrolein has been studied by many workers (see for example References 1–4), and several of the upper state vibrational modes and frequencies have been identified. From the convergence of Rydberg series, the first ionization potential has been observed at 10·10 eV,[5] this value having also been obtained by photoionization.[5] Results of electron-impact experiments have covered a range of values, the appearance potential of the molecular ion having been reported at 10·14 eV,[6] 10·25 eV[7] and 10·34 eV.[8] Calculations of the ionization potential have been reported,[9] and a value of 10·1 eV obtained. Recent calculations[10] of the orbital energies and population analysis of acrolein place the energy of the oxygen nonbonding orbital between those of the two occupied π orbitals. However, the low resolution photoelectron spectrum[11] indicated that the first ionization related to electron loss from the nonbonding orbital, as was suggested by Walsh.[12] This is confirmed by the high resolution spectrum (Figure 9.1 and 9.2).

The first band, with its 0←0 component at 10·11 eV, points to the excitation, in the ionic ground-state, of a vibration with frequency 1050 cm^{-1}. This is difficult to assign without information from deuterated derivatives, because several of the vibrational modes have very similar frequencies, and there has been no general agreement on the assignment of the vibrational frequencies.[2,4,13]

The second spectral band, (Figure 9.2 adiabatic I.P. 10·93 eV), relating to loss of an electron from the higher of the two occupied π orbitals, shows that there are at least two vibrations excited, with frequencies of 1370 and 970 cm^{-1}, and from the asymmetry of the 0←0 component and the smearing out of the higher vibrational peaks, there may also be a third mode with a frequency of about 500 cm^{-1} (cf. the upper π band of the butadiene spectrum, Figure 6.11). Frequencies of these magnitudes have been reported in the 3865 Å system of the absorption spectrum[2,4], which was designated as an $\pi^* \leftarrow n$ transition. Of the frequencies observed in the ion, the easiest to assign is 500 cm^{-1}, which probably is that of skeletal deformation, which has a molecular frequency of 570 cm^{-1}. The 970 cm^{-1} frequency is in accordance with Brand's interpretation[2] of the I.R. spectrum in that he assigns the C—C stretching mode v_{11} to a frequency of 913 cm^{-1}. This should be increased in the ion, since the upper π orbital ($2a''$) is C=C and C=O bonding, but C—C antibonding. (However, Harris[13] assigns a frequency of 1158 cm^{-1} to this mode.) Two further modes would be expected to be excited, viz. the C=O (v_5) and C=C (v_6) stretching modes, the molecular frequencies of which are quite close (1723 and 1625 cm^{-1}, respectively). It may then be that the 1370 cm^{-1} spacing is in fact the mean of two spacings, with a difference of 100–150 cm^{-1} in magnitude, in view of the fact that the peaks of the main vibrational series are broader than those in the corresponding band of the butadiene spectrum.

2. DERIVATIVES OF ACROLEIN

The position of the lower energy 'π band' in the acrolein spectrum cannot be decided *a priori*, but some further information can be obtained from the spectra of some of its derivatives.[14] The effect of methyl substitution is expected to lower the energy for π ionization relative to that for σ ionization (cf. HCN and CH$_3$CN). From the spectrum of crotonaldehyde (Figures 9.2 and 9.3), it is apparent that the I.P for the upper π orbital (10·36 eV) is lowered to a value closer to that of the nonbonding orbital (I.P. = 9·86 eV) than is the case in acrolein. In addition, the edge of the third band at 12·4 eV is at a lower I.P. than that in the acrolein spectrum ($\sim$13·2 eV). Furthermore, in trimethyl acrolein (Figure 9.6), this is lowered even further, to about 11·2 eV; and the upper π orbital is now accidentally degenerate with the nonbonding orbital. These trends point to the third band being associated with the lower π orbital in the acrolein derivatives, and hence presumably in acrolein itself.

The intensity of the first band in the spectra of the chloroacroleins (Figures 9.7–9.10) suggests that the bands due to the upper π and nonbonding orbitals are again superimposed. This can be brought about by the chlorine atom lowering the ionization energy of the π orbital ($+M$) and raising that of the nonbonding orbital ($-I$).

The 12·1 eV band can be assigned to the chlorine 'nonbonding' orbital, which has a higher energy than in vinyl chloride, ascribable to an electron withdrawing effect in the —CHO group. The 13·0 eV band is probably due to the lower π orbital.

Similar trends can be observed in comparing the spectra of crotonaldehyde and α-chlorocrotonaldehyde (Figures 9.2, 9.5).

3. GLYOXAL

This molecule is isoelectronic with acrolein, an important difference in electronic structures being that an orbital which is essentially CH_2 bonding in acrolein (a') becomes the 'in-phase' combination of oxygen $2p$ orbitals in glyoxal (b_u).

The interpretation of the photoelectron spectrum (Figures 9.11–9.13) is aided by comparison with the molecular orbitals of formaldehyde.[15] The four highest occupied orbitals of glyoxal might be expected to be the non-bonding out-of-phase and in-phase combinations of oxygen $2p$ orbitals, ($7a_g$) and ($6b_u$), and the two π-type orbitals ($1b_g$) having a node between the carbons, and ($1a_u$), with no nodes. Comparison with the spectra of iso-electronic molecules (see Figure 9.14) leads to the assignment of the first two photoelectron bands, as being due to electrons from the oxygen $7a_g$ and $6b_u$ orbitals. The difference in energy of the two nonbonding orbitals is thus approximately 1·6 eV, which can be compared with the separation of just over 0·3 eV between the two bands in *para* benzoquinone (Figure 9.15) which can be given the same assignment. The second distinct band (Figure 9.12) indicates that in the first excited state of the ion, the C=O symmetric stretching mode v_2 (a_g) and the symmetric skeletal deformation v_5 (a_g) are excited with frequencies of 1610 and 400 cm^{-1}, reduced from their molecular frequencies[13] of 1745 and 550 cm^{-1} respectively. In addition, a mode with a frequency 970 cm^{-1} may also be weakly allowed for changes of one quantum, and can be assigned to the C—C stretching mode v_4 (1060 cm^{-1} in the molecule).

The band at adiabatic I.P. 13·85 eV (Figure 9.13), which is assigned to the upper π orbital ($1b_g$) consists of a vibrational progression in v_2, the peak separation corresponding to a frequency of 1360 cm^{-1}. The greater frequency reduction of this mode in the second excited ionic state, compared with that in the first, is consistent with the greater C=O bonding character of the orbital.

The 15–16·5 eV region of the spectrum contains at least two overlapping bands.

The 17 eV band, indicates that frequencies of 1410 cm^{-1}, and 990 cm^{-1} are excited.

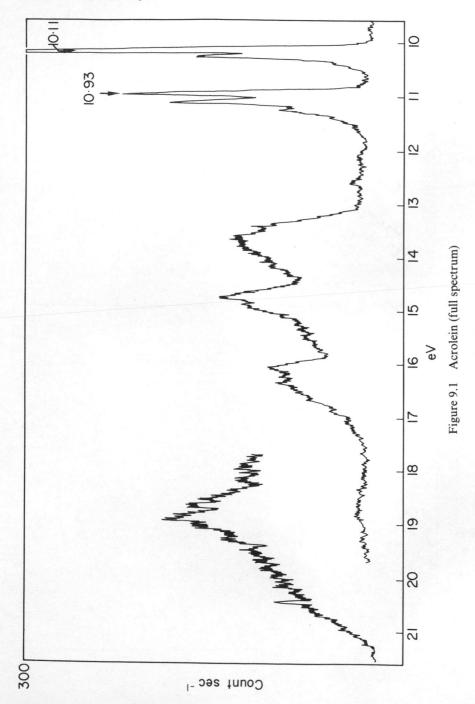

Figure 9.1 Acrolein (full spectrum)

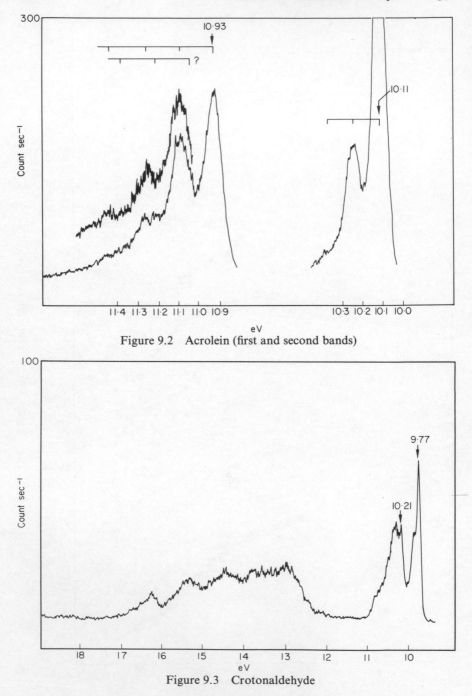

Figure 9.2   Acrolein (first and second bands)

Figure 9.3   Crotonaldehyde

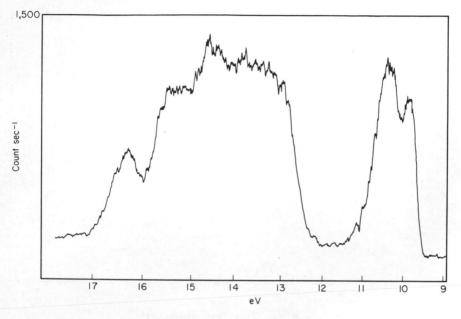

Figure 9.4   Crotonaldehyde (M)

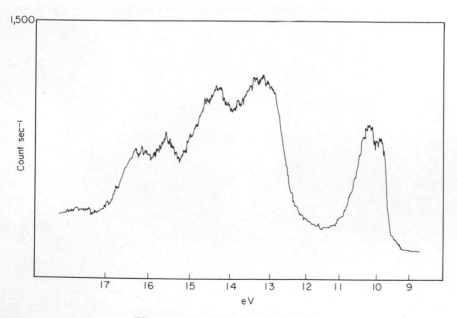

Figure 9.5   α-Methyl acrolein (M)

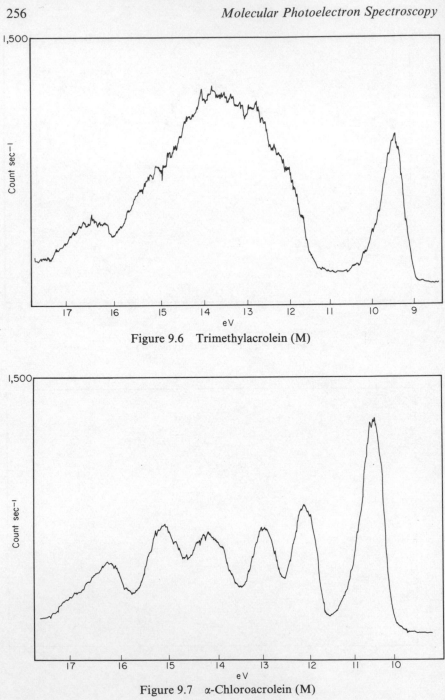

Figure 9.6   Trimethylacrolein (M)

Figure 9.7   α-Chloroacrolein (M)

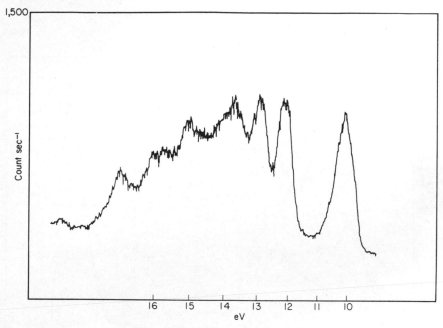

Figure 9.8 β-Chloroacrolein (M)

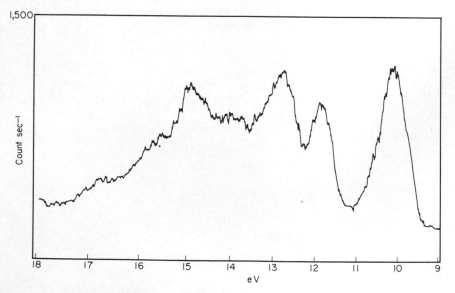

Figure 9.9 α-Chlorocrotonaldehyde (M)

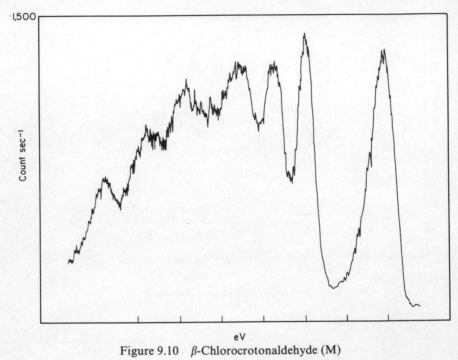

Figure 9.10 β-Chlorocrotonaldehyde (M)

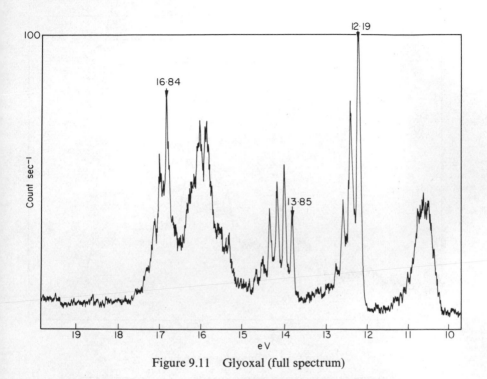

Figure 9.11   Glyoxal (full spectrum)

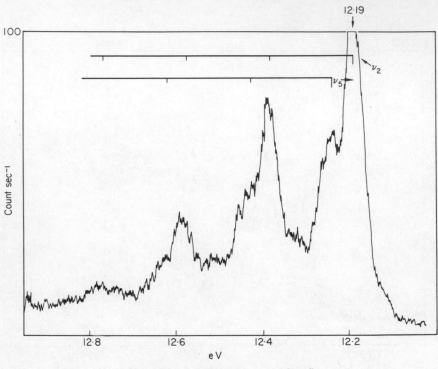

Figure 9.12   Glyoxal (second band)

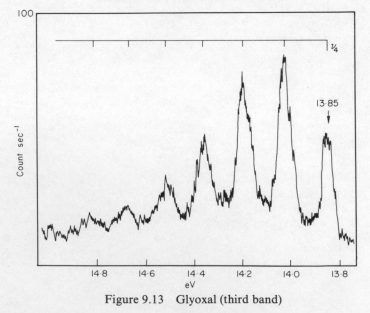

Figure 9.13   Glyoxal (third band)

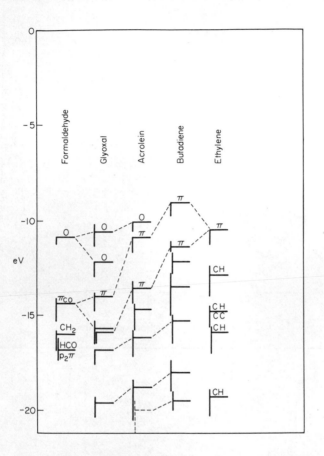

Figure 9.14   Correlation of the energy levels of the iso-π-electronic molecules of glyoxal, acrolein and butadiene showing their relationship to formaldehyde on the one hand and ethylene on the other

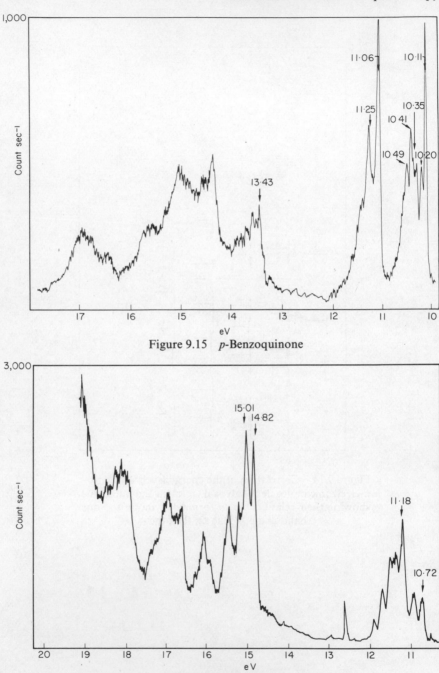

Figure 9.15   *p*-Benzoquinone

Figure 9.16   Tetrafluoro-*p*-benzoquinone [R]

REFERENCES

1. E. Eastwood and C. P. Snow, *Proc. Roy. Soc.* (*London*), *Ser. A*, **149**, 446 (1935).
2. J. C. D. Brand and D. G. Williamson, *Discussions Faraday Soc.*, **35**, 184 (1963).
3. J. M. Hollas, *Spectrochim. Acta*, **19**, 1425 (1963).
4. K. Inuzuka, *Bull. Chem. Soc. Japan*, **34**, 729 (1961).
5. K. Watanabe, *J. Chem. Phys.*, **26**, 542 (1957).
6. R. I. Reed and M. B. Thornley, *Trans. Faraday Soc.*, **54**, 949 (1958).
7. I. Omura, K. Higasi and H. Baba, *Bull. Chem. Soc. Japan*, **29**, 501, 504 (1956).
8. J. D. Morrison and A. J. C. Nicholson, *J. Chem. Phys.*, **20**, 1021 (1952).
9. F. L. Pilar, *J. Chem. Phys.*, **47**, 5375 (1967).
10. N. Jungen and H. Labhart, *Theoret. Chim. Acta*, **9**, 345 (1968).
11. D. W. Turner, *Advan. Phys. Org. Chem.*, **4**, 31 (1966).
12. A. D. Walsh, *Trans. Faraday Soc.*, **41**, 498 (1945).
13. R. K. Harris, *Spectrochim. Acta*, **20**, 1129 (1964).
14. D. P. May, Ph.D. Thesis, London University, 1966.
15. A. D. Baker, C. Baker, C. R. Brundle and D. W. Turner, *Intern. J. Mass Spectr. Ion Phys.*, **1**, 285 (1968).

CHAPTER 10

Benzene

1. THE ELECTRONIC STRUCTURE OF BENZENE

Benzene is one of the the the simplest molecules to contain multicentre π orbitals. Hückel[1] demonstrated that when the six p_z atomic orbitals of the carbon atoms constituting the benzene ring combine, six π molecular orbitals result, two pairs of these being doubly degenerate (Figure 10.1). The stability of the molecule, and the properties of its lower excited states are just two aspects of benzene chemistry on which data concerning the energies of these π molecular orbitals can throw light.

The π orbitals which are occupied in the ground state of the molecule are e_{1_g} (doubly degenerate) and a_{2_u}. The e_{1_g} orbitals have long been recognized as the highest energy orbital of the benzene molecule, but there has been some controversy as to whether a_{2_u} is the next highest energy orbital.

In 1935 Price and Wood[2] found Rydberg series for benzene in the far u.v. Later work in this region[3, 4] confirmed Price and Wood's measurements—Rydberg series converging to limits at 9·247 eV, 11·48 eV and 16·84 eV being reported. The original He 584 photoelectron experiments on benzene, carried out with a coaxial grid energy analyser, detected a number of electronic states of $C_6H_6^+$, including those found from Rydberg series. No states of the $C_6H_6^+$ ion between 9·3 eV and 11·5 eV were found however, and it was concluded that these two values were the first and second ionization potentials, and corresponded to the ejection of electrons from the upper and lower occupied π orbitals.[5] Later experiments[6] (cf. Chapter 11) on substituted benzenes, with a magnetic focusing photoelectron spectrometer of low resolution led to a similar conclusion. The validity of this conclusion was suspected when the results of MO calculations on benzene were published.[7-11] Both the semiempirical treatments of Clark and others and the *ab initio* treatment of Moskowitz and coworkers using Gaussian orbitals ($x^c y^m z^n e^{-\zeta r^2}$) seemed to indicate that the energy of the $1a_{2u}$ π orbital was lower than some of the π orbitals. In Table 10.1 the results of the various MO calculations are collected with the energetic ordering of orbitals proposed by Lindholm and

Jonsson.[12] In 1968, Momigny and collaborators at Liège University[13] attempted to reconcile the theoretical predictions with experimental observations by proposing that breaks at 10·3 and 10·8 eV in photoionization efficiency curves[13, 14] for the production of $C_6H_6^+$ ions from benzene represented two σ ionization potentials. Momigny concluded that the reason that conventional photoelectron spectroscopy found no σ ionization potentials between 9·3 eV and 11·5 eV was that the ionization cross-sections from such σ orbitals were too small when 584 Å photons were used.

The full helium 584 Å high resolution photoelectron spectrum of benzene (127° analyser) is shown in Figure 10.2. Figures 10.3 to 10.7 show various bands in the spectra of C_6H_6 and C_6D_6 to an expanded ionization energy scale. It can be clearly seen that there are no bands at 10·3 cV or 10·8 eV. We estimate that the ionization from the two σ levels Momigny proposed would have to have cross-sections of less than 0·5% that of ionization from the upper π orbital (I.P. 9·25 eV) to escape detection by us in the 9·3–11·5 eV region.

It was suggested in Momigny's original paper that the use of argon instead of helium resonance radiation to obtain a photoelectron spectrum of benzene might reveal extra bands owing to the enhanced ionization cross-sections for certain orbitals when the incident photon wavelength is increased. Natalis and his associates[15] indeed claim to have found extra bands for I.P.'s 10·3 and 10·8 eV using argon resonance-line spectroscopy. This implied that the 584 Å line sometimes fails to give all the accessible electronic states of the molecular ion. The argon resonance-line high-resolution photoelectron spectrum of benzene, however (Figure 10.7), does *not* indicate the existence of any ionization processes undetectable at 584 Å. It seems however that though the argon resonance lamp[16] normally emits lines at 1067 and 1048 Å (11·62 and 11·84 eV), the Lyman alpha line of hydrogen at 1216 Å (10·22 eV), and a nitrogen resonance line at 1134 Å are very difficult to remove unless strict precautions are taken, and are usually quite intense. Thus all the features on the argon resonance line spectrum of benzene in Figure 10.8 can be accounted for entirely as a result of the ionization by all the lines just mentioned of those orbital levels already revealed by the He 584 Å source. The overall form of Collin and Natalis's spectrum is similar in general shape and therefore it seems probable that their 'extra I.P.s' may be accounted for in a similar fashion.

Evidence from substituted benzenes (see below) again points to there being no benzene σ ionization potentials lower than 11 eV. This is summarized with other aspects of the controversy in Reference 17.

We now consider in some detail the bands in the benzene spectrum which show vibrational fine structure—the bands involved being the ones with their onsets near 9·3 eV, 11·5 eV, and 16·9 eV.

2. FINE STRUCTURE IN THE PHOTOELECTRON SPECTRUM OF BENZENE

(A) First band—onset 9·25 eV (see Figure 10.3)

A series of fairly sharp peaks tails off into a region of rather obscure structure above 9·5 eV. That an underlying σ level is responsible for this obscurity can be ruled out since one would then expect the effects of certain substituents (—CF_3 especially) to separate such a σ band from the π band. No separation of this sort is in fact observed (see Chapter 11). It may be instead that the Jahn–Teller splitting expected to occur in the doubly degenerate electronic ground state of the benzene cation is responsible for the disappearance of simple fine structure above 9·5 eV.

When Lyman alpha (1216 Å, 10·2 eV) radiation is used to obtain the spectrum, only ionization from the π $1e_{1g}$ orbital level is observed—electrons being ejected with energies less than 1 eV. The resolution of the electrostatic analyser is better at low electron energies, and the definition of the fine structure in the lowest I.P. band is more distinct in the Lyman alpha spectrum. What appeared at first sight to be a simple progression is revealed in fact to be a series of ' doublets '—the values for the two vibrational frequencies responsible being 0·07(5) eV and 0·11(5) eV, i.e. $\sim 610 \text{ cm}^{-1}$ and $\sim 930 \text{ cm}^{-1}$.

The 930 cm^{-1} value can be ascribed to the totally symmetrical 'ring-breathing' mode, v_2, of the C_6 ring (see Figure 10.9). This mode of vibration has also been identified in many of the Rydberg bands of benzene.[18] The second vibration (610 cm^{-1}) is not so easily assigned. It may relate to the 690 cm^{-1} progressions found in Rydberg bands below 1700 Å. Wilkinson[3] originally took these as corresponding to $2v_{18}$ but Liehr and Moffitt[22] found objections to this and assigned the value to $1v_{18}$ [$v_{18} = 606 \text{ cm}^{-1}$ in the ground state of the molecule]. Herzberg[18] has also postulated that v_{17} may be responsible. That a non-totally symmetric vibration is excited in 1 quantum units is good evidence that Jahn–Teller forces are in operation.

There is little change in the appearance of the fine structure associated with the 9·3 eV band on going from C_6H_6 to C_6D_6.

(B) Second band—onset 11·49 eV (see Figures 10.4 and 10.5)

The fine structure associated with the second band is complex and an unambiguous assignment is not possible from an examination of the spectrum of benzene alone. However when the spectra of benzene and benzene-d_6 are examined in conjunction, the problem of analysis is simplified, since deuteration reduces the frequency of C—H vibrational modes, but leaves C—C modes almost unchanged.

It is clear from a comparison of the two spectra that on going from C_6H_6

to C_6D_6, the peak B becomes broader and the peak C narrower. These observations imply that, in benzene, part of peak C included a contribution from a peak in a C—H vibrational progression (v_1) is the totally symmetric C—H stretch). On deuteration, the frequency of this vibration decreases, and the result in the photoelectron spectrum is that the CH 'part' of peak C becomes a CD 'part' of peak B.

The spacing AB in the benzene spectrum corresponds most probably to the frequency of the totally symmetric C—C ring-breathing vibrational mode (v_2), since it is only slightly affected on deuteration.

The question which now presents itself is how does the peak 'C', or rather, the part of it which is not affected on deuteration, fit in with the analysis of the vibrational modes excited upon the second ionization. We have come to the conclusion that there is, in fact, no obvious way in which it can be fitted in without invoking the excitation of non-totally symmetric modes, and thus it probably marks the adiabatic third ionization potential of benzene. The vibrational fine structure associated with this ionization process is then also interpretable on the basis of the excitation of totally symmetric C—H and C—C stretches.

(C) Band with onset near 16·8 eV (see Figures 10.6 and 10.7)

Once again the analysis for the fine structure associated with this band is clarified when the spectrum of C_6H_6 is compared with that of C_6D_6. Originally we believed the structure to be a simple progression in v_2, but the shape of the Franck–Condon envelope of peaks did not appear to be compatible with this—the peak 'D' being too intense with respect to peak 'C'. The 16·8 eV band of the C_6D_6 spectrum shows that v_1 is also excited weakly for part of peak 'D' in the benzene spectrum becomes displaced towards peak 'C' on going to C_6D_6.

We are now in a position to discuss these spectral features in relation to the occupied orbitals of benzene. The first point which should be made is that the use of Koopmans' Theorem may not be a suitable approximation for the comparison of the I.P.'s and orbital energies of π electrons and σ electrons. That is to say, the reorientation energies following π and σ ionizations may be very different. If this were so it would mean that the ordering of ionization potentials given by photoelectron spectroscopy need not be the same as the ordering of the orbital energies. It is possible to eliminate the necessity of taking reorientation effects into consideration by using the equation

$$\text{I.P.} = \text{Total Energy Ion} - \text{Total Energy Molecule}$$

These 'total energies' *are not the Hartree–Fock energies*, however, since they include electron correlation energies and relativistic energy terms.[19] The use of the normal Hartree–Fock method therefore cannot be expected to predict

Table 10.1   Calculated orbital energies for C_6H_6 (eV)

Orbitals placed in order suggested by Lindholm et al.	$2a_{1g}^{2}$	$2e_{1u}^{4}$	$2e_{2g}^{4}$	$3a_{1g}^{2}$	$2b_{1u}^{2}$	$1b_{2u}^{2}$	$3e_{1u}^{4}$	$1a_{2u}^{2}$	$3e_{2g}^{4}$	$1e_{1g}^{4}$
Orbital type (after Lindholm et al.) Ref. 12			weak C↔C weak C—H	weak C—C strong C—H	C↔C strong C—H	strong C—C	strong C—H	strong C—C π	weak C—C weak C—H	weak C—C π
Extended Hückel calculations (Hoffman) 7	29·6	25·8	19·9	16·6	16·6	14·3	14·6	14·5	12·8	12·8
Gaussian LCAO-SCF calculations (Moskowitz et al.) 8	26·9	24·2	19·4	15·5	15·3	12·2	13·0	12·3	10·2	7·8
Semi-empirical SCF calculations (Clark et al.) 9	31·2	24·8	20·1	19·6	14·3	13·8	13·0	15·2	9·8	9·4
Semi-empirical SCF calculations (Newton et al.)* 10								15·6		8·3

* The calculations of Newton and coworkers also show unspecified σ levels at 12·10, 15·27, 15·48, 15·79 and 21·02 eV. Dewar and coworkers have also arrived at values of 10·2 eV, 11·5, 12·7, 12·8, 13·5, 15·7, 16·1 and 19·0 eV for the first eight ionization potentials of benzene. They also believe that the second I.P. relates to a σ electron.

ionization potentials as no account of either correlation or relativistic energies is taken. Because of the complications of reorientation, relativistic and correlation energies it is perhaps not too surprising that the orderings of ionization potentials found by photoelectron spectroscopy, and the ordering of Hartree–Fock determined orbital energies are sometimes at variance. This would seem to be the case in benzene.

Secondly, it will be noted that the MO calculations all agree with one another in predicting that eight orbital levels will fall in the ionization potential range below 21, four being doubly degenerate. A total of ten bands with onsets at 9·3, 11·5, 11·7, 12·2, 13·8, 14·7, 15·4, 16·8, 18·3 and 19·9 eV can be picked out however. Further if one supposes the broad region of electron emission starting at 13·8 eV and extending to 14·7 eV to consist of two overlapping bands, then at least 11 orbital levels are indicated or 12 if the 9·3–9·5 eV band is interpreted as a Jahn–Teller doublet. This corresponds to the maximum number of bands expected above $-21·22$ eV if all degenerate ionic states are split by Jahn–Teller forces. Clearly on this count alone there cannot be two sigma I.P.'s at 10·3 eV and 10·8 eV.

Thirdly, we must note that where there are overlapping bands in the photoelectron spectrum it is not possible to derive adiabatic ionization potentials (e.g. 12·1 eV is clearly too high for the fourth adiabatic I.P.). This is an important point to bear in mind when working out Rydberg transitions as Lindholm has done to support his assignments for the ordering of the orbital energy-levels.

A slight adjustment of the structure proposed by Lindholm involving the assignment of the 11·5 eV I.P. to the $1a_{2u}$ (π_1) level, and the recognition of the 11·7 eV value as a separate I.P. still allows the prediction of Rydberg transitions in fair agreement with Lassettre's electron impact energy loss data,[20] although Lindholm himself has pointed out[21] that this electronic structure does not fully account for the strong peak at 11 eV in the spectrum of Lassettre.

Further, if our vibrational analysis for the 11·5 and 11·7 eV bands is correct, then the extremely small values of the C—H vibrational frequencies in the ions as compared with the molecules are surprising and merit comment. The implication is that both the orbitals with these ionization potentials are strongly C—H bonding as well as being somewhat C—C bonding, and it is difficult to reconcile this with our giving the 11·5 eV ionization potential to the $1a_{2u}\pi$ level. Clearly there are still a number of problems still to be sorted out relating to both the interpretation of the photoelectron spectrum and the electronic structure of benzene.

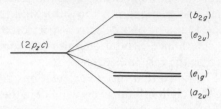

$(2p_z c)$ (b_{2g})
               (e_{2u})
               (e_{1g})
               (a_{2u})

a_{2u}, π_1 e_{1g}, π_3 e_{1g}, π_2

(a)

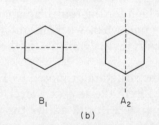

B_1 A_2

(b)

Figure 10.1   The ordering and degeneracy of the π-type molecular orbitals of benzene

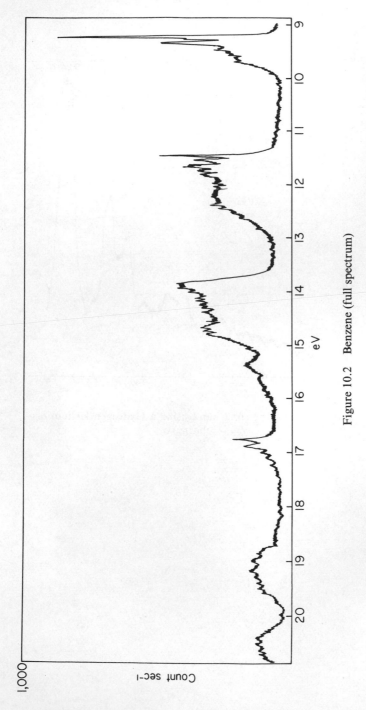

Figure 10.2   Benzene (full spectrum)

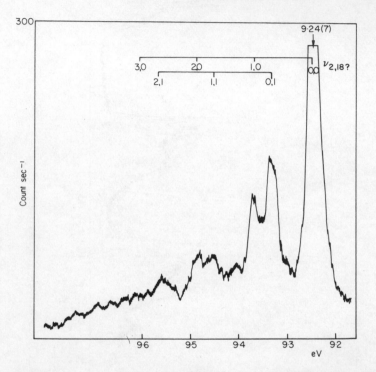

Figure 10.3   Benzene (first band using a Hydrogen/Helium discharge)

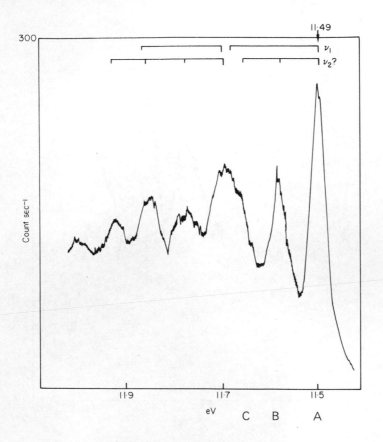

Figure 10.4　Benzene (second band)

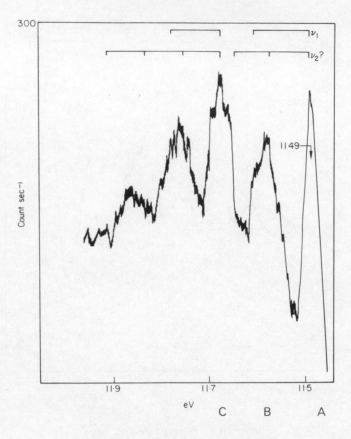

Figure 10.5   Benzene-d_6 (second band)

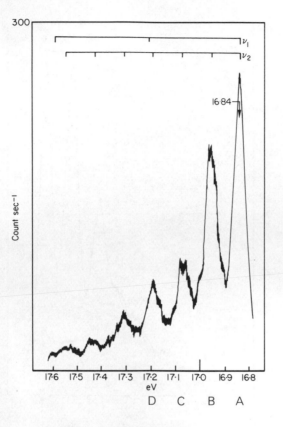

Figure 10.6   Benzene ('16·8 eV' band)

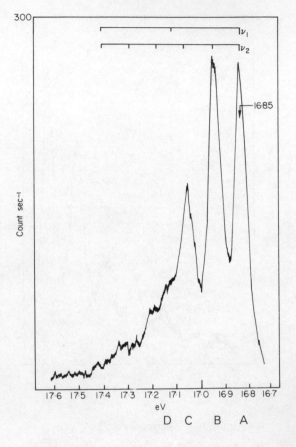

Figure 10.7   Benzene-d_6 ('16·8 eV' band)

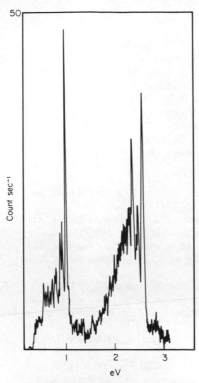

Figure 10.8 Benzene photoelectron spectrum excited by argon discharge (see text)

ν_1 A_{1g} 3061 cm^{-1}

ν_2 A_{1g} 992 cm^{-1}

ν_{16} E_{2g} 1584 cm^{-1}

ν_{17} E_{2g} 1178 cm^{-1}

ν_{18} E_{2g} 606 cm^{-1}

ν_{20} E_{2u} 404 cm^{-1}

Figure 10.9 The normal modes of vibration of the benzene molecule with relevance to the photoelectron spectrum (see text)

REFERENCES

1. E. Hückel, *Z. Physik*, **60**, 423 (1930).
2. W. C. Price and R. W. Wood, *J. Chem. Phys.*, **3**, 439 (1935).
3. P. G. Wilkinson, *Can. J. Phys.*, **34**, 596 (1956).
4. M. A. El-Sayed, M. Kasha and V. Tanaka, *J. Chem. Phys.*, **34**, 334 (1961).
5. M. I. Al-Joboury and D. W. Turner, *J. Chem. Soc.*, **1964**, 4434.
6. A. D. Baker, D. P. May and D. W. Turner, *J. Chem. Soc.*, *B*, **1968**, 22.
7. R. Hoffman, *J. Chem. Phys.*, **39**, 1397 (1963).
8. J. M. Schulman and J. W. Moskowitz, *J. Chem. Phys.*, **43**, 3287 (1965).
9. P. A. Clark and J. L. Ragle, *J. Chem. Phys.*, **46**, 4235 (1967).
10. M. D. Newton, F. P. Boer and W. N. Linscomb, *J. Am. Chem. Soc.*, **88**, 2353 (1966).
11. M. J. S. Dewar and G. Klopman, *J. Am. Chem. Soc.*, **89**, 3089 (1967); cf. also *Tetrahedron Letters*, **25**, 2341 (1967).
12. B. Jonsson, E. Lindholm, *Chem. Phys. Letters*, **1**, 501 (1967).
13. J. Momigny, C. Foffart and L. D'Or, *Int. J. Mass. Spectr. Ion Phys.*, **1**, 53 (1968).
14. V. H. Dibeler and R. M. Reese, *J. Res. Nat. Bur. Std.*, *A*, **68**, 409 (1964).
15. P. Natalis, J. E. Collin and J. Momigny, *Int. J. Mass. Spec. Ion Phys.*, **1**, 327 (1968).
16. T. N. Radwan, *Ph.D. Thesis*, University of London, 1966; A. D. Baker, *Ph.D. Thesis*, University of London, 1968.
17. A. D. Baker, C. R. Brundle and D. W. Turner, *Int. J. Mass Spec. Ion Phys.*, **1**, 443 (1968).
18. G. Herzberg, *Molecular Spectra and Molecular Structure*, Part III, van Nostrand, Princeton, 1966.
19. J. G. Stamper, *Ann. Rep. Chem. Soc.*, **64**, 24 (1967).
20. E. N. Lassettre, Unpublished work; see Reference 12.
21. E. Lindholm, Private communication to the authors.
22. A. D. Liehr and W. Moffitt, *J. Chem. Phys.*, **25**, 1074 (1956).

CHAPTER 11

Carbocyclic Aromatic Compounds

1. SUBSTITUTED BENZENE DERIVATIVES

The changes which occur in the binding energies of particular electrons in a molecule when substituent atoms or groupings are introduced are frequently mirrored in the different types of chemical reactions undergone by structurally related materials. This is particularly apparent in substituted benzene derivatives. It was an extensive study of the differences in the rates of aromatic electrophilic substitution reactions which first led Ingold[1] to believe that substituent effects could be subdivided into Inductive (I) and Mesomeric (M) components. Results from such kinetic experiments were interpreted essentially in valence-bond theory terms on the basis of a spatial flow of electrons. For example, a rate-enhancing substituent which led to *ortho–para* substitution was always found to be one which could contribute a p electron pair to the aromatic system and thus transfer electronic charge to the *ortho* or *para* reaction sites.

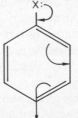

The groups NMe_2 and OMe with loosely bound $2p$ electrons clearly belonged in this category. Conversely, reaction-retarding substituents which led to *meta* substitution were easily categorized as uniquely electron-withdrawing. In certain cases it was found necessary to recognize that the two types of behaviour could act in opposition. For example, there were groups which, though rate-depressing, afforded *ortho–para* orientated substitution products.

A semiquantitative rationalization of substituent effects in organic chemistry has been provided by the Hammett linear-free-energy relationship[2, 3, 4]

$$\log (k/k_0) = \rho\sigma$$

in which k and k_0 are equilibrium or rate constants for the substituted and unsubstituted compounds respectively, σ is a constant of the substituent, and ρ another constant of some reaction at a particular atomic site within the molecule concerned. The constants σ for a number of substituents provide a quantitative scale upon which the electronic properties of the substituents can be assessed. Elaboration of this simple concept has resulted in the introduction of modified constants ($\sigma^0, \sigma^+, \sigma^-$) which cater for particular situations according to the electronic nature of the reaction.

It has been suggested that separation of the overall effect of substituents (σ) into inductive (σ_I) and resonance (σ_R) contributions (where $\sigma = \sigma_R + \sigma_I$) is made possible by estimating σ_I from aliphatic reactions,[5] or, in the case of fluorine, from ^{19}F chemical shifts in *meta* substituted fluorobenzenes.[6] The means of making this separation has, however, been criticized by Dewar and Marchand.[7] On the basis of optical spectroscopy, a further subdivision of the inductive effect (I) into π and σ components (I_π and I_σ) has been made. Here, the σ inductive effect (I_σ) is taken as arising from changes in the electronegativity of the carbon atom to which the substituent is attached, whilst the π inductive effect relates to the direct repulsions between aromatic π electrons and substituent p electrons.[8] Finally, Dewar and his associates have shown that in some cases I_σ (understood here as a bond-relayed effect) can be replaced by a 'through space' polarization of the aromatic electrons under the influence of the dipolar electrostatic field in the vicinity of strongly electron-withdrawing substituents, especially CF_3.[7]

The changes in energy which accompany the spatial distortion of the π electron distributions brought about by a substituent are calculable by MO theory. Such calculations suffer, however, from uncertainties as to the values of certain important parameters, notably the values for deeper energy levels in the substituent group and especially in the ordering of the energy levels in benzene itself. The largest π level perturbations can be expected to arise when occupied p or π levels of the substituent are very close in energy to the π levels of the benzene ring. As pointed out in Chapter 10, there has been considerable argument in recent years as to the energy associated with values of these benzene ring π electrons.

Calculations have indicated overlap in the energy range between π and σ levels, and certain experimental data has also been interpreted in this way (cf. Chapter 10). What is unambiguous, however, is that two (degenerate) π orbitals are the highest occupied of all the benzene orbitals. It is well known that the degeneracy of the upper π levels is removed by substitution, as also are the degeneracies between the members of the four pairs of carbon L-shell σ orbitals, but the extent of the level splittings which will result in particular cases could not be measured with confidence without the use of photoelectron spectroscopy. It will become apparent from the spectra of benzene derivatives shown in this chapter that the energy separation between the π levels found for different substituents allows a quantitative scale to be drawn up which reflects the mesomeric electron-releasing powers of the substituent groupings. Previous attempts to measure substituent effects by using just the first ionization potential (measured from photon or electron impact ionization efficiency curves, or from Rydberg series limits) had of course to ignore the π level splittings because of lack of data. There had also been the uncertainty as to whether in a substituted benzene the most easily removed electron characterized by the first ionization potential is a benzene π electron and not a lone-pair electron of the substituent, as might be the case in (for example) benzaldehyde.

The photoelectron spectra of some benzene derivatives are shown in Figures 11.1–11.48. The spectra marked 'M' were obtained by using a magnetic sector analyser.[9, 10, 11] Other spectra were obtained using the higher-resolution electrostatic-sector apparatus described in Chapter 2.

(A) Unifying Features in the Spectra of Benzene Derivatives

Certain common features can be traced throughout the photoelectron spectra of benzene derivatives. The first band of the benzene spectrum ($\sim 9\cdot3$ eV) has a counterpart in the spectra of substituted benzenes, but in several instances it is broadened (e.g. fluorobenzene, toluene) or split into two components (e.g. chlorobenzene, phenol, aniline) and is frequently shifted to higher or lower ionization energy. In a later section (p. 282) these effects are discussed in terms of the substituents removing the degeneracy of the upper π levels, and changing their ionization potentials by mesomeric and inductive interactions.

The gap of nearly 2 eV between the first two bands in the spectrum of benzene is frequently occupied by one or more noticeably sharper peaks in the spectra of monosubstituted benzenes (e.g. in the bromobenzene and aniline spectra). These peaks can be discussed in terms of the ionization of electrons largely localized in substituent lone-pair type orbitals (see p. 285). Bands believed to be derived from the ionization of electrons in some other π-type orbitals of substituents are considered briefly in Reference 12.

The second, third, and fourth bands in the benzene spectrum (corresponding to adiabatic I.P.'s of $11\cdot49$, $11\cdot69$ and $\sim12\cdot1$ eV) are barely identifiable as separate bands. In the photoelectron spectra of substituted benzenes, there is frequently a merging of the three bands so that a broad 'hump' is obtained in the I.P. region at about $11\cdot5$–$13\cdot0$ eV.

The broad and almost featureless band in the photoelectron spectrum of benzene covering the range $13\cdot5$–$16\cdot0$ eV is probably best interpreted as being entirely due to the ionization of electrons from sigma orbitals. Little useful information can be obtained at present from this region of the spectrum owing to the broadness of the band. Further, the picture is complicated even more in substituted benzenes because of the removal of the degeneracies of some of the σ levels, and the introduction of addition σ electrons due to the substituent.

A band near 17 eV is a feature common to the spectra of several substituted benzenes. The effects of substituents on the position in the spectrum of this band are discussed in section (E) and these effects indicate that perhaps the best interpretation for this band is to assign its presence as due to ionization from a symmetrical, predominantly C—C σ bonding orbital.

In this chapter, and elsewhere, we have attached to various bands in the spectra the names of different 'types' of orbitals, e.g. benzene π orbital, halogen lone-pair orbital. These names, although not rigorously correct on the basis of MO theory, are useful for the characterization of common spectral features which can be rationalized in terms of a tendency towards localization in specific molecular or atomic sites of the electron before ejection. These expressions should not therefore be taken as implying complete electron localization.

(B) The Effects of Substituents on the First Band (9·25 eV) of the Benzene Photoelectron Spectrum

Monosubstituted benzenes

It has been pointed out already that the twofold degeneracy of $^2E_{1g}$ ground state of the benzene positive ion is expected to be removed by the presence of substituents. This is evident from an examination of the symmetries of the highest occupied π orbitals in benzene (cf. Figure 10.1). At the point of substitution, one of the orbitals (B type; π_3) has its maximum electron-density whilst the other (A_2 type; π_2) has a node.

That such a loss of degeneracy occurs is clearly demonstrated in the photoelectron spectra of many substituted benzenes. Thus in the spectra of chlorobenzene and bromobenzene, the lowest-energy ionization bands are clearly double, whilst in iodobenzene, the splitting of the π_3/π_2 orbital energy levels appears to be so large that the π_2 photoelectron band overlaps a spectral band believed to relate to the ionization of an iodine $5p$ electron. The relative orders of the π_3/π_2 splittings in the halogenobenzenes are in the same order as the $+M$ effects associated with the halogens, and thus appear to reflect the abilities of the 'lone-pair' electrons to donate electronic charge to the benzene ring, affecting only the B_1-type orbital.

Photoelectron spectra of phenol and aniline derivatives also show similar trends. In each spectrum for instance distinct bands (relating to the two lowest ionization potentials) can be picked out corresponding to the π_3 and π_2 (B_1 and A_2) components of the $(e_{1g})^4$ benzene level. Furthermore, in aniline and its N-alkyl derivatives (Figure 11.17), the second I.P. (π_2) remains nearly constant, whilst the first (π_3) changes greatly from compound to compound, again demonstrating the ability of only the B_1-type orbital to accept electrons donated by the mesomeric effect from the substituent.

The increased splittings in the O- and N-alkyl derivatives are doubtless associated with the release of electrons from the alkyl group to the oxygen atom facilitating further mesomeric release of electrons into the benzene ring.

It will be noticed from the spectra of certain monosubstituted benzene-derivatives (e.g. $C_6H_5.CF_3$, Figure 11·6, $C_6H_5.CN$, Figure 11·10, $C_6H_5.CHO$, Figure 11·11) that there appears to be no or very little splitting of the π_3 and π_2 orbital energy-levels. In all these cases, however, the π_3/π_2 band is moved by several tenths of a volt to higher ionization-energy than the band of lowest I.P. in benzene. The substituents producing these effects are in each case ones which are classically regarded as being efficient at removing charge from the benzene ring by the $-M$ effect (e.g. —CN, —NO) or by the $-I$ effect (e.g. —CF$_3$). The spectra of fluoroalkyl phenyl ethers also show no π_3/π_2 splittings (in addition c.f. $C_6H_5N(CF_3)_2$, Figure 11·20). Here it would seem that the introduction of the fluorine atoms alters the nature of the oxygen $2p$ lone-pair of electrons to such an extent that they are no longer effective in conjugating with the ring π electrons.

The π electron perturbations which all these $-M$ and $-I$ substituents produce are thus not confined to the B_1 orbital. It seems therefore that there must be either efficient transmittance of the inductive effect to ortho positions or a 'through-space' polarization[7] both of which perturb the A_2 orbital.

Para disubstituted benzenes

Introduction of two substituents, A and B, into the 1,4 positions of the benzene ring often leads to the separation of the π_3 and π_2 photoelectron bands being nearly equal to the sum of the separations in the monosubstituted compounds, C_6H_5A and C_6H_5B. This additivity of substituent effects is again an expected consequence of the symmetries of the π_3 and π_2 orbitals —the former having its maximum electron-density at positions 1 and 4, the latter having nodes at both points of substitution.

Fairly large divergences, however, from the additivity phenomenon are apparent in a number of cases suggestive of some kind of 'saturation' [e.g. p-bromoanisole, (Figure 11.39), p-methyl-N,N-dimethylaniline (Figure 11.42), and p-dimethoxybenzene (Figure 11.43)]. In such molecules, one of the substituents acting alone is capable of causing a large perturbation of the π electron density in the benzene ring, and this seemingly prevents the second substituent from exerting as large an effect as might otherwise be expected.

There are two very obvious difficulties frequently encountered in the interpretation of the variations arising from structure changes in the positions of photoelectron spectral bands. Firstly bands in similar positions in photo-electron spectra may not relate to equivalent MO's in different molecules. Secondly it is not always clear whether to use 'vertical' or 'adiabatic' I.P.'s if fine structure is resolved, and if it is not, what is to be used then?

Misinterpretations of the first kind can usually be avoided by careful corre-lation of the spectra of a large number of compounds. In connection with the second point we may argue that vertical I.P.'s are perhaps the most useful to theoretical chemists since their calculations for ions normally assume the same geometry as the molecule. High resolution photoelectron spectra for benzene derivatives suggest that in several cases, e.g. for weakly bonding halogen 'lone pairs' and some π orbitals, the vertical ionization potential is clearly defined by the most intense vibrational peak in a series. Frequently it appears that the vertical I.P.'s are equal to adiabatic I.P.'s in such cases. The distinction between 'vertical' and 'non-vertical' ionization is only really clear, however, when it relates to diatomic molecules. In all other cases, the complexity of the potential surfaces could lead to a situation in which vertical ionization, because it can generate vibration in a number of different modes of similar *but not quite identical* energy might appear to have a smaller prob-ability per unit energy range than for example the adiabatic ionization simply because for the latter all the ionization probability is concentrated in one vibronic component. For benzene especially the high resolution photo-electron spectrum may over-emphasize the adiabatic ionization to the ionic ground-state.

For certain purposes then it may be more appropriate to employ the 'centres' of photoelectron bands to obtain 'mean' I.P.'s. When we measure the splittings of the upper benzene E_{1g} π levels in its substitution products better 'additivity' in predicting E_{1g} splittings in p disubstituted derivatives is obtained using these values than splittings obtained using either specifically adiabatic or vertical I.P.'s (cf. Reference 12).

Ortho and meta disubstituted benzenes

From the relatively small range of *ortho* and *meta* disubstituted benzenes so far examined (Figures 11.45–11.50) it seems to be general that where both substituents contribute to the moving apart of the π_2 and π_3 levels, the actual splittings are greater in the cases of the p disubstituted compounds than in the corresponding *ortho* and *meta* derivatives. Also the effect is usually more marked in *meta* compounds than *ortho*.

Pentafluoro substituted benzene derivatives

It is evident from the spectra of all the C_6F_5X compounds (Figures 11.51–11.57) that the π–π splittings in each case are considerably less than in the corresponding C_6H_5X compounds. In fact, the only compound examined so far which clearly shows a doubling of the bond with lowest I.P. is $C_6F_5NH_2$ (Figure 11.57), where the $\pi_3-\pi_2$ splitting is ~ 0.8 eV (compare 1·1 eV for $C_6H_5NH_2$).

The spectrum of C_6F_6 (Figure 11.58) is especially interesting since the high symmetry of the molecule results in the vibrational fine structure of some of the bands being especially simple—either the totally symmetric C—C stretching mode is excited, or this together with the totally symmetric C—F stretch. The first adiabatic I.P. of C_6F_6 as given by photoelectron spectroscopy is 9·97 eV, in agreement with Price's previous work.[13] We have not been able to discover two bands reported by Clark and Frost[14] in a photoelectron spectrum obtained with a spherical grid analyser, and counted as π (I.P.'s 11·12 and 11·64).

(C) Calculations on the A_2 and B_1 π Ionization Potentials of Benzene Derivatives

Theoretically calculated values of the energy differences between the A_2 and B_1 type π orbitals in benzene derivatives have been obtained by Caldow. [15,16]

The earlier calculations[15] on the lowest ionization potentials of the monohalogeno benzenes indicated only a small splitting of the degeneracy of the benzene E_{1g} levels.

In view of the much larger splittings indicated by photoelectron spectroscopy[12,14] (cf. the spectra shown in this chapter), Caldow has now concluded that a large inductive parameter must be allowed for in his modified ω-technique (see Streitwieser[17]) for the calculation of ionization potentials. Some of Caldow's new results[16] indicate one or two anomalies in the photoelectron results by Clark and Frost[12] for polyfluorobenzenes with a spherical-grid energy analyser (cf. the photoelectron results for C_6F_6 obtained with spherical-grid and electrostatic-sector analyser).

(D) Bands Associated with Substituent Lone-Pairs of Electrons

The photoelectron spectra of halogenobenzenes and some other substituted benzenes contain relatively sharp peaks not resembling any feature in the spectrum of benzene. These peaks are best interpreted in terms of the ionization of electrons largely localized in substituent lone-pair type orbitals. Chlorobenzene, for example, has in its photoelectron-spectrum narrow bands corresponding to adiabatic I.P.'s of 11·3 and 11·7 eV—these being interpreted as arising from the ionization of chlorine $3p$ lone-pair electrons. This interpretation is supported by the following facts: (i) All chloro compounds so far examined by photoelectron spectroscopy have distinctive bands in the

I.P. region 10·5–12·5 eV (see Chapter 8). (ii) The shapes of the Franck–Condon envelopes are indicative that the orbitals involved are nonbonding or very weakly bonding.

By similar reasoning it can be deduced that the narrow bands in the spectra of bromobenzene (adiabatic I.P.'s 10·6(1) and 11·1(8) eV) and iodobenzene (adiabatic I.P.'s 9·6(4) and 10·4(5) eV), and the sharp band in the spectrum of fluorobenzene near 13·9 eV are the results of ionizations of bromine $4p$, iodine $5p$, and fluorine $2p$ lone-pair electrons.

It is perhaps rather puzzling at first sight why there should be two halogen lone-pair bands in the spectra of C_6H_5Cl, C_6H_5Br, and C_6H_5I (Figures 11·2–11·4). The reason appears to be that the p_x and p_y halogen orbitals interact differently with the benzene π system. The p_y orbital, that is the one which is perpendicular to the C–Halogen bond and the plane of the benzene ring, would certainly be expected to overlap more effectively with the benzene orbitals than p_x. The consequences of this would be that the 'p_y' electrons had a higher I.P. than the 'p_x', and also that the p_y orbital lost more of its nonbonding character than the p_x. All the halogenobenzene spectra containing two halogen bands quite clearly show that of the halogen 'lone-pair' bands the one at lower I.P. is narrower (lower bonding character), and furthermore the vibrational spacings, corresponding most probably to the stretching of the C–Halogen bond, seem to be larger in the halogen bands appearing at lower I.P.

For the monohalogenobenzenes, the order of separations of the p_x and p_y orbitals is F (unsplit) < Cl < Br < I. The amount of splitting of the halogen bands, and also the positions of the bands in the spectra are very sensitive to the presence of substituent groupings in the benzene ring. Thus, in the spectrum of p-bromobenzotrifluoride (Figure 11.36), the p_x and p_y bromine lone pair bands are actually further apart than in bromobenzene itself, whereas the amount of splitting is appreciably diminished in p-bromotoluene (Figure 11.35), p-bromophenol (Figure 11.38), and in pentafluorobromobenzene (Figure 11.56). The bromine peaks are also at much higher ionization energy in C_6F_5Br than in C_6H_5Br. Relatively sharp 'lone-pair bands' associated with substituents other than halogens (i.e. N or O) are also evident in the spectra of aniline, substituted anilines, and some phenol derivatives. Such bands, however, are broader than the halogen lone pair bands, indicating some incorporation of the lone pair into the benzene-ring framework.

In phenol itself (Figure 11.15), it appears that the I.P. of the oxygen '$2p$ lone pair' (11·3 eV) is about the same as in that of one of the benzene orbitals since no separate sharp peak due to the lone pair can be seen in the spectrum.

Again, the positions occupied by the N or O 'lone-pair' bands in the spectrum depend critically upon the nature of the substituents in the ring. It is possible that 'chemical shifts' of this sort may prove to be useful in qualitative analysis.

(E) The Effects of Substituents on Benzene Orbitals other than the A_2 and B_1 (π_2 and π_3)

The presence of substituents always causes more bands to appear in the photoelectron spectrum than are present in that of benzene. Consequently it is not always easy or even possible to trace the changes in all the orbital ionization energies which occur on going from benzene to its substitution products. It is possible, however, to make some limited measurements on the shifts incurred by substitution on the bands present at 11·5 eV and 16·8 eV in benzene.

Since both these bands show fine structure in the benzene spectrum, it was felt that one of the bands probably corresponded to ionization from the lowest π level ($1a_{2u}$, π_1) and that the differing effects of substituents on the two bands might reveal which of the bands were σ and which π. In Table 11.1 values for the ionization potentials derived from these bands in the spectra of a number of substituted benzenes are given. The effects of halogen substituents on the 16·8 eV band merit comment. Cl, Br and I substitutions do not significantly alter the position of the band, but the effect for —Br and —I is to lower I.P., whereas —F substitution causes the band to move to higher ionization energy by about 0·6 eV. That —F substitution produces an increase in this I.P. and —Br and —I a slight decrease perhaps suggests that here the predominant effect is an inductive one, and this in turn would imply that the 16·8 eV benzene band relates to a σ-type orbital. Effects produced by substituents on the 11·5 eV benzene band are difficult to rationalize. This is perhaps because there are many orbitals in this region, so that one can never be sure that the value recorded for bands in similar positions in spectra of different compounds in fact relate to the same orbitals. Again the halogenobenzenes form an interesting group. The band corresponding to the 11·5 eV band in benzene is at 12·2 eV in chlorobenzene, 11·8 eV in bromobenzene, 11·8 eV in fluorobenzene, and 11·3 eV in iodobenzene. The further presence of a chlorine atom in the p position of all these monohalogen compounds leads to an additional increase of 0·5 eV in the I.P. value associated with this band.

These results are not altogether inconsistent with the assignment of the 11·5 eV benzene band to the $1a_{2u}$ (π_1) orbital. Mesomeric interaction between the π_1 orbital and the chlorine 'lone pair' would be expected to raise the π_1 orbital I.P. in contrast to the effect on the π_3 I.P. (lowered) because the π_1 I.P. is higher and the π_3 I.P. lower than the chlorine $3p$ lone-pair I.P.

Also the larger changes of the 11·5 eV benzene I.P. induced by chlorine substitution in comparison with fluorine substitution would not be expected if the orbital involved were σ in type.

Table 11.1   Benzene derivative I.P.'s which correspond to the 11·5 eV and 16·8 eV I.P.'s of benzene

Compound	I.P.'s		Fig.
C_6H_6	11·5	16·8	36·1
C_6H_5Br	11·8	16·7$_7$	36·5
C_6H_5I	11·3	16·4	36·6
C_6H_5Cl	12·2	16·9	36·4
C_6H_5F	11·8	17·3	36·3
$C_6H_5CF_3$	11·9	18·1	36·7
$C_6H_5CH_3$	11·2	16·4	36·12
$C_6H_5C[CH_3]_3$	10·5	16·5	36·14
F—⬡—F	12·2	17·1? 17·7?	36·25
F—⬡—Cl	12·3	17·4	36·26
Cl—⬡—Cl	12·6	16·9	36·27
Br—⬡—Cl	12·3	~16·7	36·35
CH_3—⬡—Cl	~11·5	~16·5	36·28
CF_3—⬡—Cl	12·4	>18	36·29

Table 11.1—*continued*

Compound	I.P.'s		Fig.
CF_3—⬡—CF_3	12·4	18·6	36·48
Br—⬡—CH_3	11·5	~16·5	36·37

The values given in this Table are derived from $0 \leftarrow 0$ vibrational components of bands, or from band onsets, and are thus approximately 'adiabatic'.

2. NAPHTHALENE AND OTHER POLYCYCLIC AROMATICS

The photoelectron spectrum of naphthalene obtained with 584 Å radiation shows three narrow low I.P. bands, two of which show resolved vibrational fine structure (Figures 11.59, 11.60, 11.61). The spectra of indene, quinoline and isoquinoline likewise contain three low energy bands similar in appearance to those of naphthalene.[18]

Since narrowness of spectral bands frequently indicates ionization from π-type orbitals it is perhaps reasonable to assign these bands to the three highest occupied π orbitals. There are indeed strong grounds for believing that the first *five* ionization potentials of naphthalene relate to the five occupied π orbitals of the molecule. Simple Hückel MO theory, for example, leads to such a prediction—Danby and Eland[18] have used the first three I.P.'s of naphthalene, indene, azulene and diphenyl to obtain the equation

$$\text{I.P.} = 6.22 + (2.77 \pm 0.15)m$$

where 2.77 ± 0.15 is β, the resonance integral and m is the HMO coefficient.

This equation then places the first five π I.P.'s of naphthalene below 13.2 eV, equal to the number of bands in the photoelectron spectrum below this value. The band with its onset at ~11.0 eV and extending to ~12.2 eV is very complex in form, however, and perhaps should be regarded as two or more overlapping bands in the same way as was the 11.5 eV band of benzene. This should then place the highest occupied σ orbitals of both molecules at about 11.7 eV.

A final interesting point regarding the applicability of simple Hückel theory is that the differences between the first and second ionization potentials of both benzene and diacetylene (see Chapter 6) accord well with the predictions of Hückel theory incorporating a value of 2.2 eV for the parameter β.

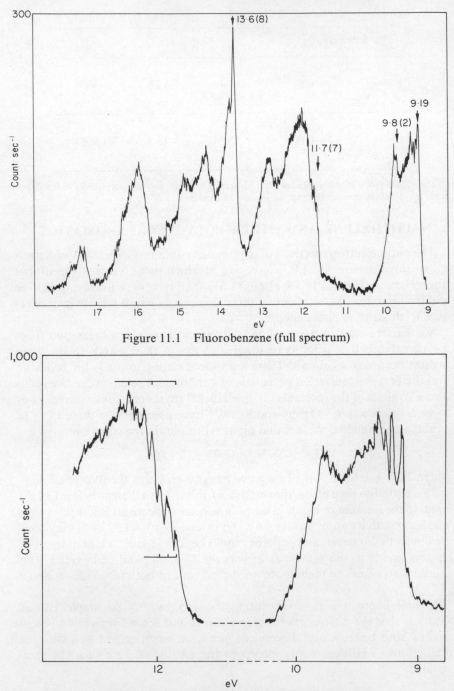

Figure 11.1   Fluorobenzene (full spectrum)

Figure 11.2   Fluorobenzene (first and second bands)

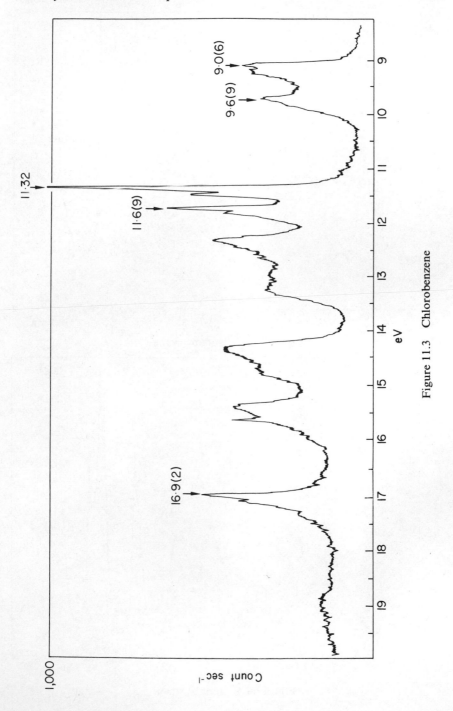

Figure 11.3 Chlorobenzene

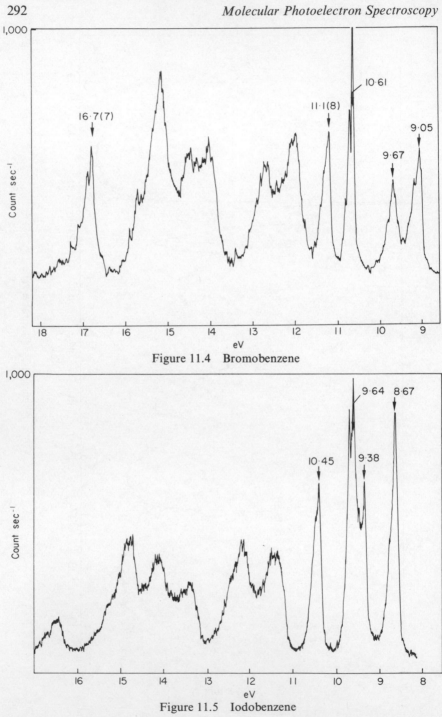

Figure 11.4   Bromobenzene

Figure 11.5   Iodobenzene

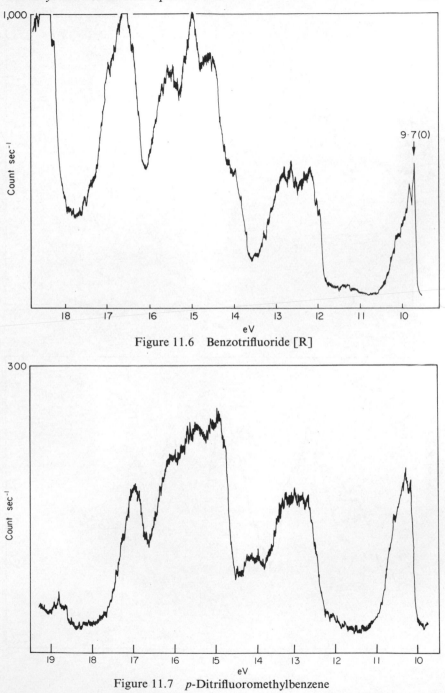

Figure 11.6   Benzotrifluoride [R]

Figure 11.7   *p*-Ditrifluoromethylbenzene

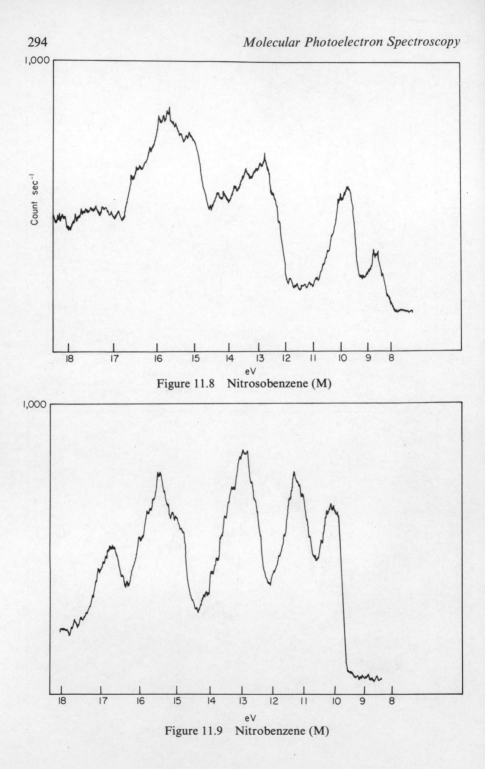

Figure 11.8   Nitrosobenzene (M)

Figure 11.9   Nitrobenzene (M)

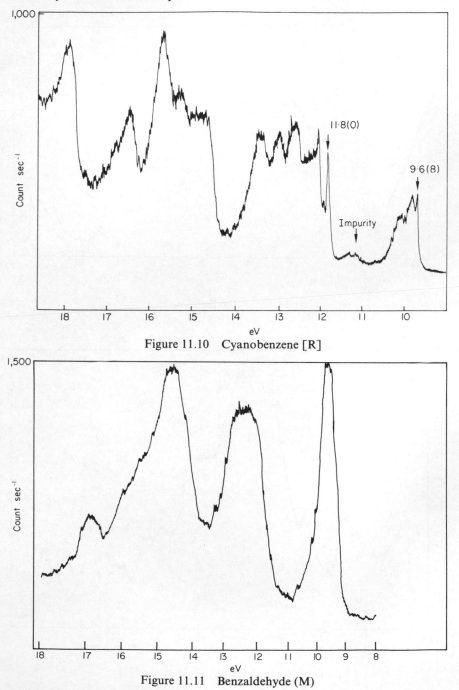

Figure 11.10   Cyanobenzene [R]

Figure 11.11   Benzaldehyde (M)

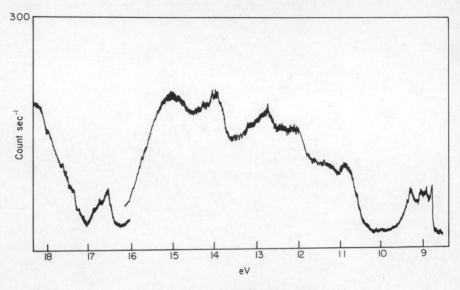

Figure 11.12   *t*-Butylbenzene [R]

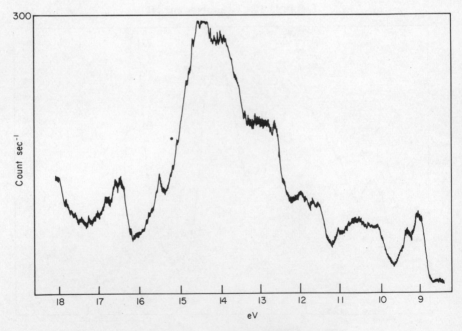

Figure 11.13   Trimethylsilylbenzene [R]

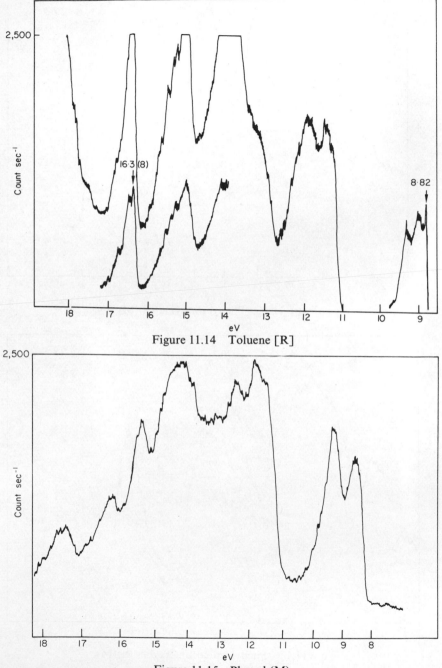

Figure 11.14   Toluene [R]

Figure 11.15   Phenol (M)

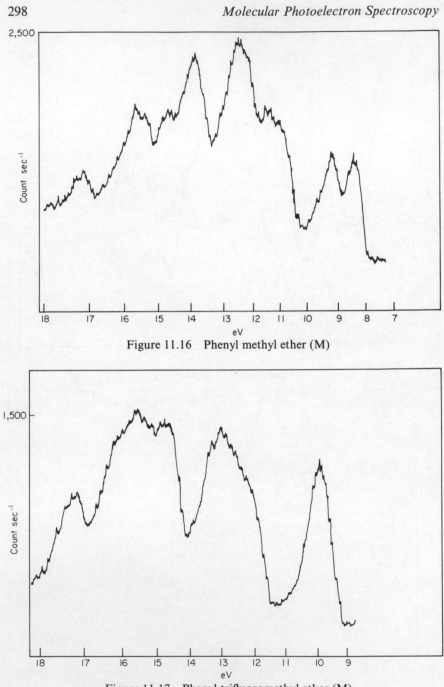

Figure 11.16   Phenyl methyl ether (M)

Figure 11.17   Phenyl trifluoromethyl ether (M)

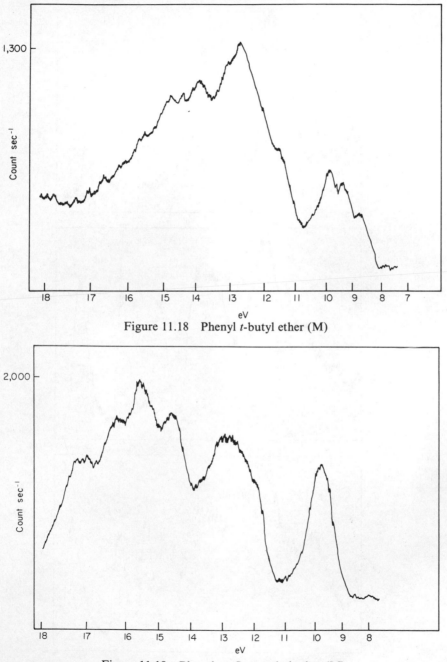

Figure 11.18   Phenyl *t*-butyl ether (M)

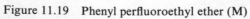

Figure 11.19   Phenyl perfluoroethyl ether (M)

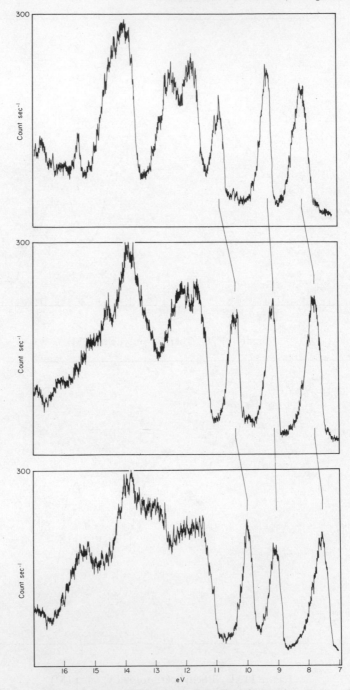

Figure 11.20   Aniline, *N*-methylaniline and *N,N*-dimethylaniline (top to bottom)

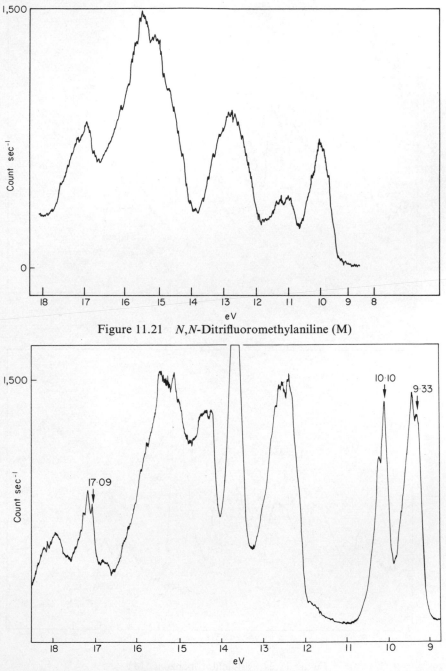

Figure 11.21   *N,N*-Ditrifluoromethylaniline (M)

Figure 11.22   *p*-Difluorobenzene

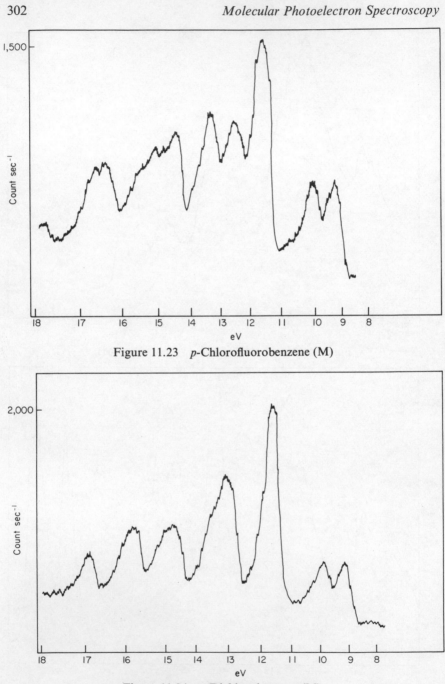

Figure 11.23   *p*-Chlorofluorobenzene (M)

Figure 11.24   *p*-Dichlorobenzene (M)

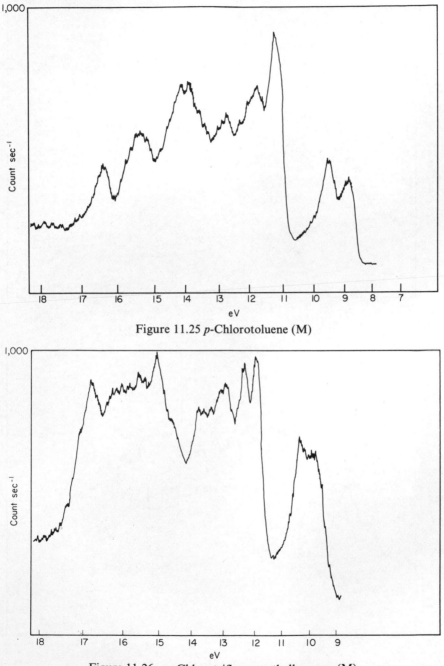

Figure 11.25 *p*-Chlorotoluene (M)

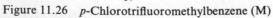

Figure 11.26 *p*-Chlorotrifluoromethylbenzene (M)

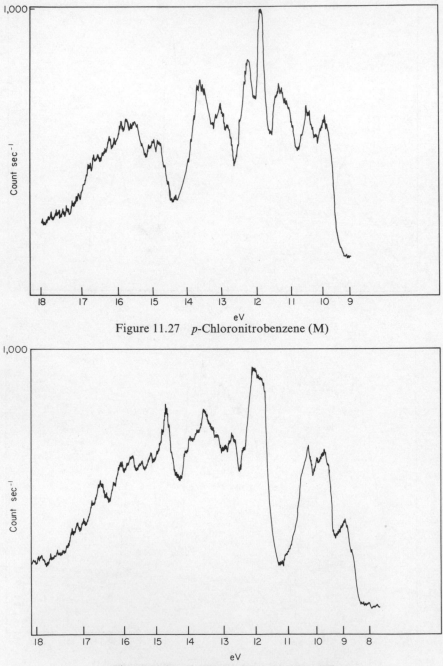

Figure 11.27   *p*-Chloronitrobenzene (M)

Figure 11.28   *p*-Chloronitrosobenzene (M)

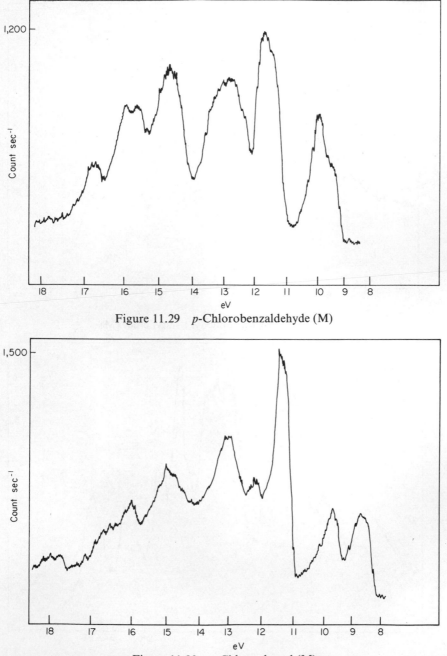

Figure 11.29   *p*-Chlorobenzaldehyde (M)

Figure 11.30   *p*-Chlorophenol (M)

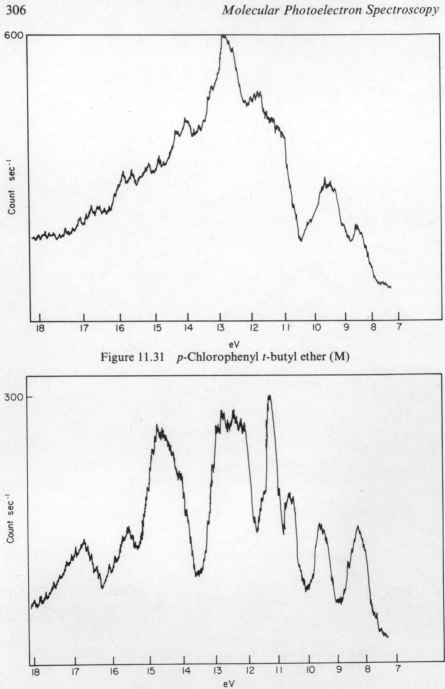

Figure 11.31 *p*-Chlorophenyl *t*-butyl ether (M)

Figure 11.32 *p*-Chloroaniline (M)

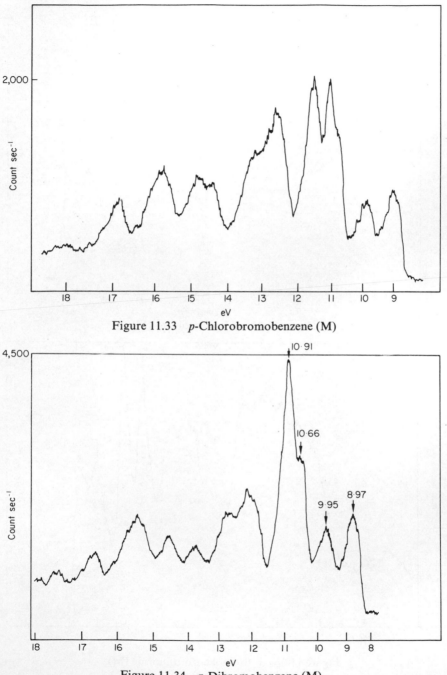

Figure 11.33   *p*-Chlorobromobenzene (M)

Figure 11.34   *p*-Dibromobenzene (M)

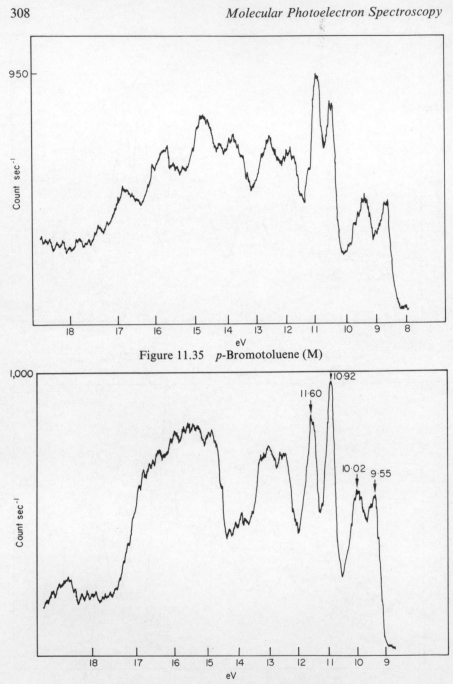

Figure 11.35   *p*-Bromotoluene (M)

Figure 11.36   *p*-Bromobenzotrifluoride (M)

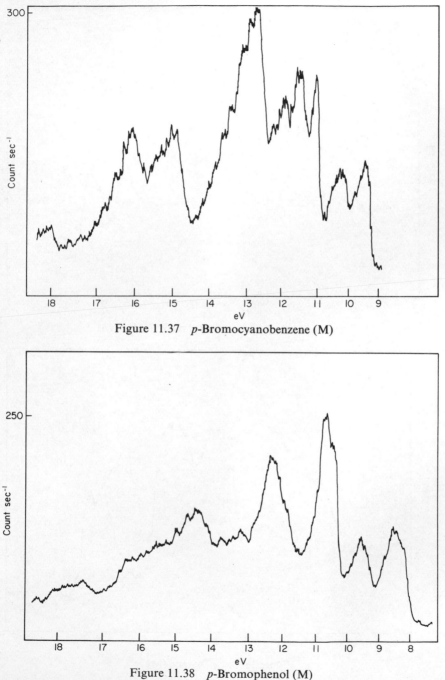

Figure 11.37 *p*-Bromocyanobenzene (M)

Figure 11.38 *p*-Bromophenol (M)

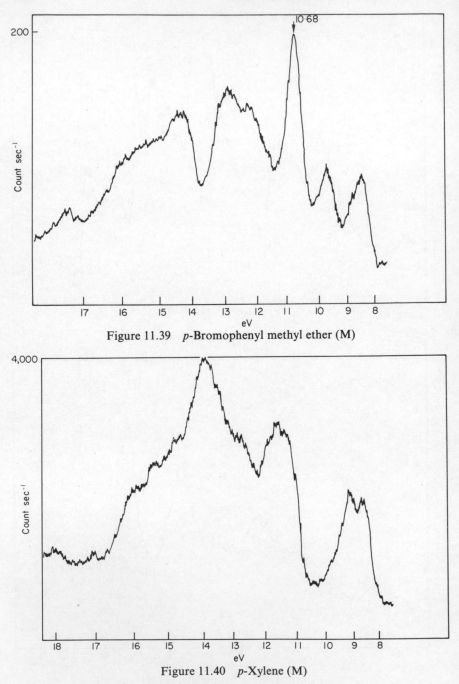

Figure 11.39   *p*-Bromophenyl methyl ether (M)

Figure 11.40   *p*-Xylene (M)

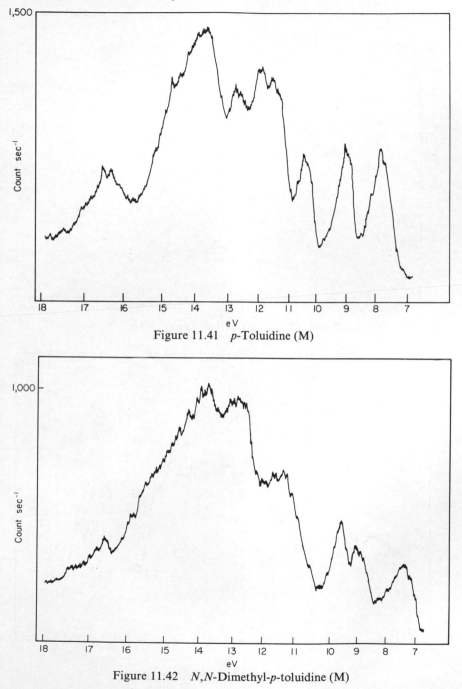

Figure 11.41 *p*-Toluidine (M)

Figure 11.42 *N,N*-Dimethyl-*p*-toluidine (M)

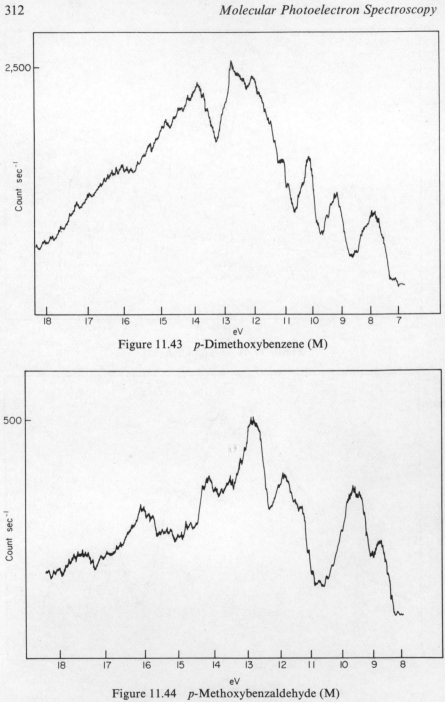

Figure 11.43   *p*-Dimethoxybenzene (M)

Figure 11.44   *p*-Methoxybenzaldehyde (M)

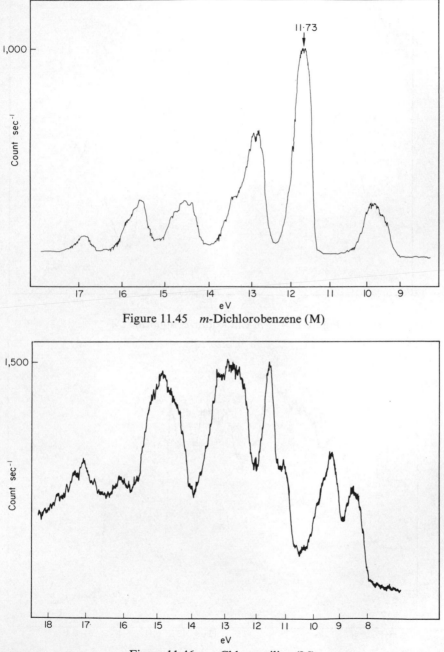

Figure 11.45   *m*-Dichlorobenzene (M)

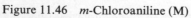

Figure 11.46   *m*-Chloroaniline (M)

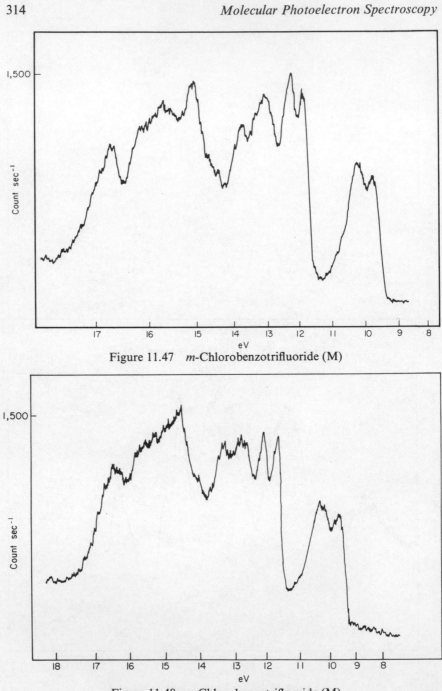

Figure 11.47   *m*-Chlorobenzotrifluoride (M)

Figure 11.48   *o*-Chlorobenzotrifluoride (M)

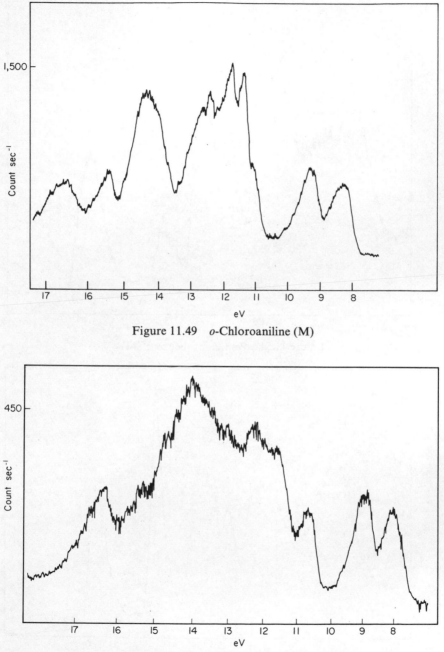

Figure 11.49   *o*-Chloroaniline (M)

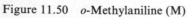

Figure 11.50   *o*-Methylaniline (M)

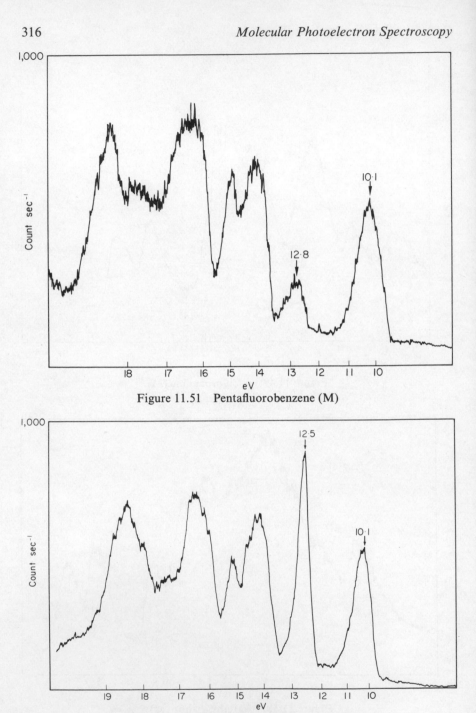

Figure 11.51   Pentafluorobenzene (M)

Figure 11.52   Pentafluorochlorobenzene (M)

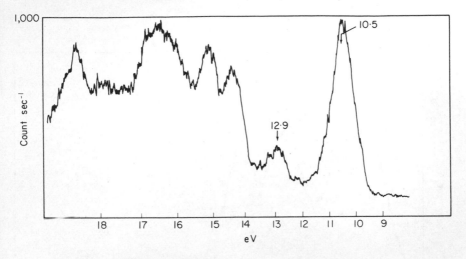

Figure 11.53   Pentafluorobenzaldehyde (M)

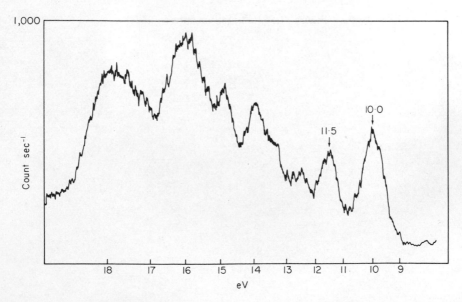

Figure 11.54   Pentafluorophenyl methyl ether (M)

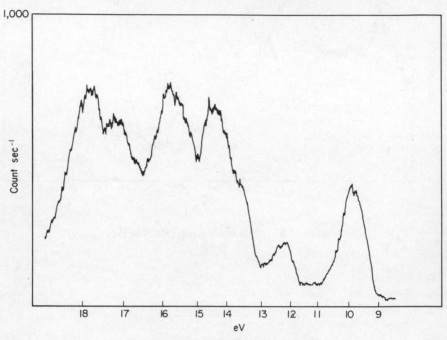

Figure 11.55   Pentafluorophenol (M)

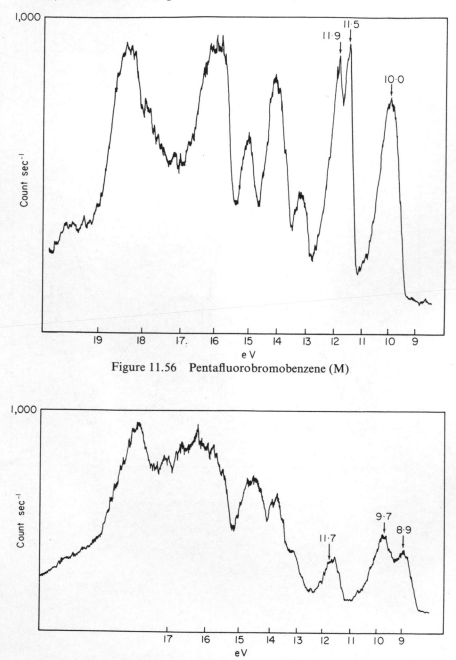

Figure 11.56 Pentafluorobromobenzene (M)

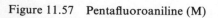

Figure 11.57 Pentafluoroaniline (M)

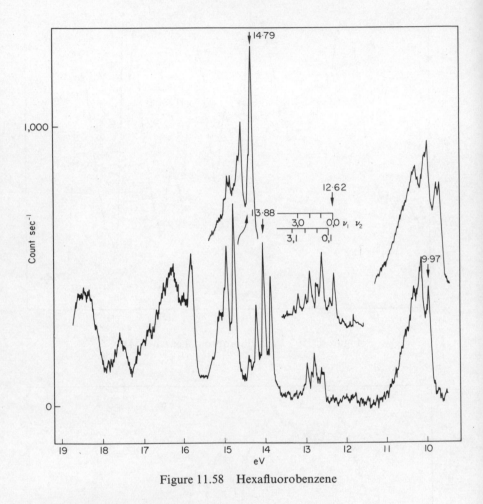

Figure 11.58   Hexafluorobenzene

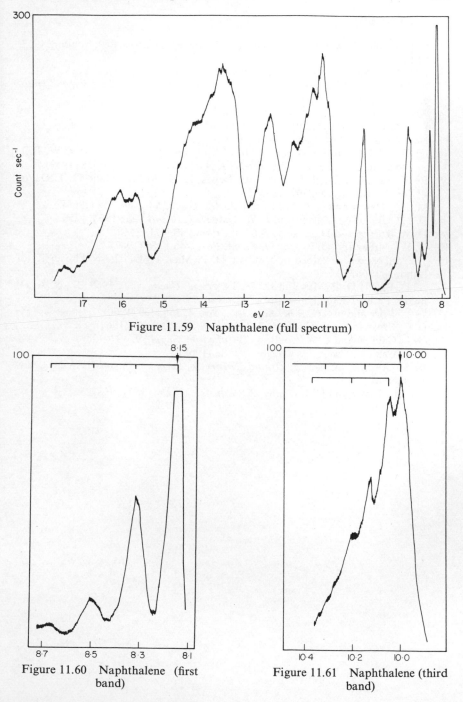

Figure 11.59   Naphthalene (full spectrum)

Figure 11.60   Naphthalene (first band)

Figure 11.61   Naphthalene (third band)

REFERENCES

1. C. K. Ingold, *Structure and Mechanism in Organic Chemistry*, Bell, London, 1953, Ch. 6.
2. L. P. Hammett, *Physical Organic Chemistry*, McGraw-Hill, New York, 1940.
3. P. B. D. de la Mare and J. H. Ridd, *Aromatic Substitution: Nitration and Halogenation*, Butterworths, London, 1959.
4. 'Quantitative Evaluation of Substituent Effects by Electronic Spectroscopy' by J. N. Murrell, and 'Quantitative Aspects of Aromatic Substitution' by R. O. C. Normal in *Roy. Inst. Chem. (London), Lectures*, **1963**, 2.
5. Cf. R. W. Taft and I. C. Lewis, *J. Am. Chem. Soc.*, **81**, 5343 (1959); R. W. Taft, S. Ehrenson, I. C. Lewis and E. Glick, *J. Am. Chem. Soc.*, **81**, 5352 (1959).
6. R. W. Taft, E. Price, I. K. Fox, I. C. Lewis, K. K. Anderson and G. T. Davis, *J. Am. Chem. Soc.*, **85**, 709 (1963).
7. M. J. S. Dewar and A. P. Marchand, *J. Am. Chem. Soc.*, **88**, 354 (1966).
8. D. J. Clark, J. N. Murrell and J. M. Tedder, *J. Chem. Soc.*, **1963**, 1250.
9. D. W. Turner and D. P. May, *J. Chem. Phys.*, **45**, 471 (1966).
10. D. W. Turner and D. P. May, *J. Chem. Phys.*, **46** 1156 (1967).
11. This analyser is described in detail by D. P. May, Ph.D. Thesis, University of London, 1966.
12. A. D. Baker, D. P. May and D. W. Turner, *J. Chem. Soc.*, **1968**, 22; cf. A. D. Baker, Ph.D. Thesis, University of London, 1968.
13. W. C. Price and others, *Proc. Roy. Soc., Ser. A*, **258**, 459 (1960).
14. D. C. Frost and I. D. Clark, *J. Amer. Chem. Soc.*, **89** 244 (1967).
15. G. L. Caldow and C. A. Coulson, *Tetrahedron, Suppl.*, **7**, 127–138.
16. G. L. Caldow, *Chem. Phys. Letters*, **2**, 88 (1968).
17. A. Streitweiser, *Molecular Orbital Theory for Organic Chemists*, Wiley, New York, 1961.
18. J. H. D. Eland and C. J. Danby, *Z. Naturforsch*, **23a**, 355 (1968).

CHAPTER 12

Heterocyclic Aromatic Compounds

1. INTRODUCTION

Heterocyclic aromatic compounds fall into two main classes—those in which the heteroatom lone-pair electrons are incorporated into the aromatic π system e.g. furan, and those in which they are not, for example pyridine.

The relative orderings of the π, σ and n orbitals in both these classes of compound have never been firmly established. Two main reasons may be cited. Firstly, there has been some question as to whether the highest occupied σ orbital of aromatic compounds has an energy exceeding that of the lowest π orbital (cf. Chapter 10), and secondly, for molecules in which the heteroatom $2p$ 'lone-pair' electrons are independent of the π system, there has been no sure way of finding out whether the π orbital I.P.'s are lower than the lone-pair I.P.'s

Pyridine and benzene diazines bear a close resemblance to benzene and to one another since in every case one or two of the benzene —CH functions are replaced by the isoelectronic nitrogen atom. Conflicting evidence on the ordering of the π, σ, and 'N$2p$ lone-pair' orbitals has come from both theoretical and experimental studies. Some of the evidence relating to the π/N$2p$ ordering problem was collected together and critically discussed in a paper published in 1968 by Yencha and El-Sayed.[1]

On the theoretical side, two recent series of MO calculations on pyridine have been carried out. Clementi[2] used an *ab initio* approach with contracted Gaussian orbitals, whilst Emsley[3] used the CNDO/2 method of Pople and his associates.[4] The orbital energies found by Clementi and Emsley are included in Table 12.1. Emsley's calculations indicate a complete intermingling of σ and π orbital energies. This was also suggested by the earlier work of Newton and coworkers[5] for pyridine as well as for benzene (cf. Chapter 10). Newton's work further gave the highest orbital energy as 'nitrogen lone-pair'.

The work of Clementi shows nine orbital levels of pyridine with energies in excess of $-21\cdot2$ eV. A complete description of the bonding characteristics of each orbital can be found from the appropriate population analysis charts,[2] but for the present purposes it suffices to say that the two highest energy orbitals $1a_2$ and $2b_1$ are both of the π type and the next highest energy orbital ($11a_1$) is to be described as N2p lone-pair. The lowest π orbital ($1b_1$) is calculated by Clementi as having an energy less than some of the σ orbitals. It should be noted here that comparisons between experimental I.P.'s and theoretically calculated orbital energies are subject to the same limitations mentioned in Chapter 10.

Table 12.1 Pyridine–calculated orbital energies greater than $-21\cdot2$ eV

	Clementi[2]		Emsley[3]
$1a_2$...	12·2	...	9·4
$2b_1$...	12·5	...	10·9
$11a_1$...	12·7	...	11·4
$7b_2$...	15·8	...	12·4
$1b_1$...	16·9	...	13·4
$10a_1$...	17·4	...	15·5
$6b_2$...	18·2	...	16·9
$9a_1$...	19·1	...	19·4
$5b_2$...	19·8	...	20·1
$8a_1$...	21·2		

The proper description of the 'lone pair' of electrons on the nitrogen atoms in azines has attracted much discussion, especially as to the degree of 's' character associated with the orbitals concerned. Del Bené and Jaffé[6] have carried out some calculations on this, using the CNDO method, and their results indicate a significant delocalization of the lone-pair electrons throughout the pyridine molecule. They also obtain an 's' character of about 14%. We shall return to the question of the 'lone pairs' in the diazines in a later section.

2. PYRIDINE

(A) Previous Experimental Work

Experimental studies on the ionization of pyridine and its substituted derivatives have not been very extensive. El-Sayed and coworkers[7] have identified Rydberg series converging to 9·266 eV, about 10·3 eV and about 11·56 eV, these three limits being interpreted as representing ionization potentials of the outermost π orbital (species a_2), of the nitrogen 'lone-pair' (species a_1) and of the second π orbital (b_1) respectively. This assignment was in conflict with the earlier electron impact studies of Baba, Omura, and Higasi[8] who attempted to determine the effects of substituents on the lowest ionization potentials, and deduced that the nitrogen lone-pair orbital was the highest occupied one. A later electron impact study[9] made by Basila and Clancy however gave results more in keeping with El-Sayed's assignments than with Baba's. The frequencies of charge-transfer bands reported by Krishna and Chowdhury[10] are also more compatible with El-Sayed's assignment for the first I.P.[7]

(B) Photoelectron Spectroscopic Results

The photoelectron spectrum of pyridine is shown in Figures 12.1 and 12.2. Since the first and third bands are the only ones showing fine structure, it is not possible to give precise adiabatic ionization potentials for the other bands as the $0 \leftarrow 0$ components are not identifiable. However the values 9·26, 10·5, 12·27, 13·0, 13·8, 14·2, 15·6 and 16·9 eV can be given for band 'onsets', the 9·26 and 12·27 eV values being those derived from the $0 \leftarrow 0$ vibrational components of the first and third bands. Earlier, lower resolution experiments with a coaxial-grid photoelectron energy analyser[11,12] yielded values of 9·28, 10·54, 12·22, 13·43, 14·44, 15·49, 16·94, (19·39) and (20·14) eV, the values in parentheses being the less well-established ones. The photoelectron results for the first I.P. are confirmed by the Rydberg convergence limits given by El-Sayed (9·266 eV)[7] and the photoionization yield curves obtained by Watanabe (9·23 eV).[13]

On comparison of the pyridine photoelectron spectrum with that of benzene (see Chapter 10), it is apparent that fewer of the bands show vibrational fine structure in the former case, and even in those which do, the fine structure is poorly resolved and more complex in pattern. This is to be expected from the lower symmetry of the pyridine molecule. On passing from the D_{6h} symmetry of benzene to the C_{2v} symmetry of pyridine, the degeneracy between the members of ten pairs of the benzene vibrations is removed, each degenerate vibration splitting into two components, one symmetric and one antisymmetric with respect to the C_2 z-axis. In benzene there are only two vibrations of the A_1 class, whereas in pyridine there are ten. The fine structure on the low ionization edge of the lowest I.P. band of the pyridine spectrum is too complex to permit a complete analysis. It seems however that a vibration with a frequency of about 560 cm^{-1} may be excited, and this can probably be assigned as ν_5 which has a frequency of 992 cm^{-1} in the ground state of the molecule (see Figure 12.3).

The narrowest band in the pyridine spectrum is that at I.P. about 10·5 eV. The narrowness of the band suggests that it is probably attributable to the 'nitrogen lone-pair' (Clementi's orbital $11a_1$). It is interesting to note however that the N2p lone-pair I.P. of $CH_3CH:NC_2H_5$ is only 9·29 eV.[11] The 10·5 eV band is not as narrow as bands obtained for most nonbonding halogen lone-pair orbitals (see Chapter 8) and this perhaps bears out del Bené and Jaffé's idea of appreciable delocalization. A reasonable conclusion based on these observations is that the band with onset 9·2 eV is composed of contributions from electrons of the upper two π levels. This interpretation will henceforth be referred to as 'scheme (*i*)'. An alternative must however also be considered—this involving the first three ionization potentials being π (9·2 eV), n ($\sim$9·5 eV), and π ($\sim$10·5 eV). This interpretation—'scheme (*ii*)'—requires the substitution of a nitrogen atom into the benzene nucleus to have caused an increase of about 1·2 eV in the I.P. of the B_1-type orbital.

The band near 12·5 eV in the pyridine spectrum is the only one to exhibit fine structure which is at all well-resolved. The band has previously been tentatively assigned to the lowest π level on the basis of this.[14] A possible analysis of the fine structure associated with the 12·5 eV pyridine band is to be found in Figure 12.2, this involving the excitation of vibrations with frequencies of 560 cm^{-1} (0·07 eV) and 1650 cm^{-1} (0·205 eV). However a similar structure in the corresponding band of the benzene spectrum (which is situated 0·8 eV to lower ionization energy) has now been interpreted in terms of two closely spaced ionization potentials (cf. Chapter 10). A study of the spectrum of pyridine-d_5 would clearly help with the interpretation of the vibrational structure.

Photoelectron spectra of several substituted pyridines are shown in Figures 12.4–12.10.

The effect of methyl substitution in the α position of the ring decreases the ionization potentials of the orbitals corresponding to the 9·3 and 10·5 eV bands by about 0·2 eV, the effect on the 10·5 eV band being slightly greater. A methyl group in the γ position, on the other hand, has a much more marked effect on the 10·5 eV band than the first band. The 10·5 eV band can be deduced to have shifted to lower ionization energy by more than 0·4 eV, since in the spectra of γ-methylpyridine, there is only one broad band near 10 eV, whereas there are two in pyridine.

Substitution of a chlorine atom into the α position of the ring causes a slight increase in the 10·5 eV ionization potential of pyridine. The effect of this substitution on the first band of the pyridine spectrum is to split it into two components. A chlorine atom in the ring β position has much the same effect, but substitution into the γ position has quite different consequences. It seems here that, as in the case of γ-methylpyridine, the substituent has the effect of moving the second pyridine band (10·5 eV) to lower I.P. and the first band (9·3 eV) to higher I.P., so that the two bands merge in the spectra of the γ-substituted compounds. If the 10·5 eV level of pyridine does relate to an essentially nonbonding nitrogen orbital, as scheme (*i*) requires, it seems that its energy is highly dependent on the substituent in the ring γ position. This is a little reminiscent of effects noted previously (Chapter 10) for *para* substituted chlorobenzenes, where the energy of the chlorine 'lone-pair' orbitals has been found to vary over quite a large range as the *para* substituent is changed.

The merging of the two bands of lowest I.P. on γ substitution is also explicable in terms of scheme (*ii*). The assumption then would be that the B_1 orbital I.P. was increased in pyridine with respect to benzene owing to the replacement of a C—H group by a nitrogen atom, but was lowered again by the presence of electron-releasing substituents in the γ position of the ring. Such an effect would be expected to be limited essentially to γ substituents since the B_1 orbital has its maximum electron-density both at the nitrogen atom and at the γ position of the ring.

Two striking similarities between the spectra of substituted benzenes and substituted pyridines can be noted: (*a*) The first bands of the pyridine and benzene spectra show a tendency towards a splitting into two components when substituents are introduced. This is strong, but not conclusive, evidence in support of scheme (*i*); (*b*) In halogen substituted compounds, there are frequently two bands present ascribable to 'halogen lone-pairs'. This has been discussed in Chapter 10 in terms of the p_x and p_y components interacting to different extents with the 'aromatic' π orbitals.

3. BENZENE DIAZINES

The highest filled π orbital of pyrazine has a nodal surface passing through both nitrogen atoms. In view of this, the corresponding ionization potentials in benzene, pyridine, and pyrazine would be expected to be similar. The first ionization potentials for the three molecules are 9·25, 9·26 and 9·29 eV respectively, and this close agreement indicates that the first I.P. relates to a π orbital throughout. The second highest filled π orbital of pyrazine has its maximum electron-density in the vicinity of the nitrogen atoms. In benzene, the energy separation of the highest occupied π levels produced by *para* substituents has been found to be roughly the sum of the effects produced by either substituent alone (see Chapter 10). The separation in energy of these two π orbitals in pyrazine would therefore be expected to be approximately twice that in pyridine. The band of lowest I.P. in the pyridine spectrum contains two overlapping components, which according to interpretation (*i*) of the preceding section indicates that the two lower π levels are separated by about 0·2 eV. The maxima of the two bands of lowest I.P. in the pyrazine spectrum (Figures 12.11–12.14) are separated by between 0·5 and 0·6 eV, and thus it seems that the two highest occupied π orbitals of pyrazine may have the two lowest I.P.'s. Further evidence comes from the vibrational structure resolved on the two bands of lowest ionization energy in pyrazine since it appears (see below) that the same ring-breathing mode may be involved in both cases. If this assumption were correct, a possible extension of the arguments

advanced above would be that the two lowest I.P.'s of pyridazine and pyrimidine were also π, especially since all the diazines' spectra contain two bands
in the same I.P. region as the 10·5 eV band of the pyridine spectrum, which
is given as 'nitrogen lone-pair' by scheme (*i*) enunciated in section 2 of
this chapter. However, the assignment for pyrazine based on scheme (*ii*) for
pyridine, and involving the pyrazine bands at 9·3 eV and 11·4 eV corresponding to the two upper π levels, is also consistent with the available evidence.
Further, Yencha and El-Sayed point out that it would be a little surprising
to have a first I.P. for pyridazine lower than pyrazine, and for this reason,
these workers assign the first I.P. of the former compound to the highest
occupied nitrogen orbital. This would certainly be quite consistent with
scheme (*ii*) since it would only be necessary to assume a 'crossing-over' of
the π_3 and highest occupied *n* orbital energies in pyridazine as compared to
pyridine (see Figure 12.18). The assignment given here involving the first
pyrazine I.P. as π is in agreement with Yencha and El-Sayed's conclusion,[1]
but both our suggested orbital orderings (*i*) and (*ii*) are at variance with
Clementi's[15] and with Del Bené's and Jaffé's[6] calculations. Clementi's results
for the eight highest occupied orbitals of pyrazine are summarized in
Table 12.2. Here the orbitals $10a_1$ and $11a_1$ represent the two nitrogen lone-
pairs. His calculations show that the 'upper nitrogen (*n*) orbital' $11a_1$ has
0·687 electrons on each nitrogen atom, while the lower has 0·94—the

Table 12.2   Clementi's calculated energies for the eight
highest occupied orbitals of pyrazine[15]

$11a_1$...	11·9
$1a_2$...	12·2
$2b_1$...	13·4
$10a_1$...	14·5
$7b_2$...	16·7
$1b_1$...	17·7
$6b_2$...	18·7
$9a_1$...	19·6

remaining electrons being delocalized in each case throughout the orbitals
of the molecule, principally through the $2p_y$ orbitals of the carbon atoms α
and β to the nitrogen. This compares with the results for pyridine which give
1·41 electrons on the nitrogen atom and 0·59 electrons delocalized.

Clementi also points out that the simple assumption that the two 'nitrogen
lone-pair orbitals' are derived from positive and negative combinations of
two hybrid orbitals—Equations (*i*) and (*ii*)—cannot be expected to form the
basis of an adequate description.

$$\psi_a = \text{hy N}_1 + \text{hy N}_2 \qquad\qquad (i)$$
$$\psi_b = \text{hy N}_1 - \text{hy N}_2 \qquad\qquad (ii)$$

Del Bené and Jaffé also dismiss this idea, basing this on their calculations for pyrimidine and pyridazine. Here it will be noted that the ordering of the first four orbitals in each case is π, n, π, n in disagreement with Clementi's results, and also with our tentative interpretations of the I.P. orderings deduced from the photoelectron spectra.

The reason Del Bené and Jaffé give for rejecting the concept of positive and negative combinations of hybrid orbitals is that such a description would lead one to predict a separation of the two n orbitals which was largest in the case of 1,2-diazine (i.e., pyridazine), and this is not indicated by their calculations. Our scheme (*i*) (Figure 12.18) indicates however that such an effect may be present. The separation of the third and fourth bands (the first and second having been assigned to π orbitals) is 0·4, 9·2 and 0·7 eV in the order 1,4-, 1,3- and 1,2-diazine.

The very long series of peaks associated with the lowest ionization energy band of the pyrazine spectrum (Figure 12.14) can be interpreted in terms of the excitation of vibration of frequency 615 ± 50 cm^{-1} in the ion. Yencha and El-Sayed's photoionization efficiency curves show a vibrational fine structure associated with the lowest ionization threshold of pyrazine, but the frequency they measure, 726 cm^{-1}, is a little higher than our value. The vibrational mode excited can almost certainly be attributed to the totally symmetric ring-breathing mode v_4 (Table 12.3) which has a frequency of

Table 12.3 Vibrational modes of pyrazine mentioned in
the text, with their equivalents in pyridine.
Frequencies in cm^{-1} refer to the molecular ground-state

Pyridine		Pyrazine	
v_1	3054	v_1	3054
v_4	1583	v_2	1570
v_9	992	v_4	1015
v_{10}	605	v_5	596

1015 cm^{-1} in the molecular ground-state. The vibrational mode excited on the second lowest energy ionization of pyrazine (photoelectron band 10·1 eV) would also seem to be v_4 since it has the value $1015 \pm$ cm^{-1} in the ion, and this would not be expected to differ significantly from the value in the molecule, since the orbital involved here would appear to be virtually nonbonding.

4. FURAN

The first ionization potential of furan, relating to the uppermost of its three π molecular-orbitals, has been accurately determined by Rydberg series convergence and photoionization efficiency measurements to be $8·89 \pm 0·01$ eV (see Reference 16).

As in the cases of benzene and pyridine, MO calculations predict[17] that the lowest of the three π levels will have a lower energy than some of the σ orbitals (Table 12.4). The fact that both the furan and benzene photoelectron spectra have similarly structured bands near 17 eV, and that pyridine has a fairly sharply defined photoelectron band at this energy too is probably significant,

Table 12.4   Calculated energies of occupied MO's in furan[17]

A_1	B_2	A_2	B_1
$-52\cdot876$	$-33\cdot702$	$-10\cdot128$	$-19\cdot610$
$-37\cdot071$	$-23\cdot846$		$-12\cdot079$
$-24\cdot252$	$-14\cdot109$		
$-21\cdot574$	$-12\cdot480$		
$-12\cdot905$			
$-12\cdot184$			

and must indicate an orbital of similar bonding characteristics in each case. Weak bonding character has often been associated with π symmetry,[18-20] and as the MO calculations all predict a low lying π level, it is perhaps tempting to assign the 17 eV band to this. Very different reorientation energies following the ionization of π relative to σ electrons, or very different correlation energies, however, could possibly lessen the significance of such comparisons between I.P.'s and orbital energies.

The first three electron bands of furan show fine structure (Figures 12.16–12.18), that in the third being rather indistinct. Fine structure is also apparent in the band near 17 eV. Examination of the first band reveals that there are two modes excited in the ground state of the $C_4H_4O^+$ ion. The frequencies measured for these modes are about 1300 cm^{-1} and 1000 cm^{-1}. A positive assignment of these values is difficult because several of the furan A_1 class vibrations (Figure 12·23) have ground state frequencies between 1000 and 1500 cm^{-1}. However on the basis of the nodal properties of the highest occupied π orbital (high electron density over four-carbon-atom chain) it seems reasonable to assign the 1300 cm^{-1} value to v_3 which has a frequency of 1486 cm^{-1} in the molecule and involves a contraction along the carbon chain.

The vibration excited on the second ionization has a value of ~ 900 cm^{-1} (0·11 eV) and here the totally symmetric ring-breathing mode v_7 may be involved. The reduction from 994 cm^{-1} in the molecule is about that expected on the basis of ionization from a weakly bonding orbital. There is some evidence that a C—H mode may be weakly excited also (2900 cm^{-1}) since there is a reproducible 'shoulder' separated by this amount from the adiabatic component of the band. The structure associated with the 17 eV band approximates to a simple series with a peak-to-peak spacing of 0·12 eV (i.e. 970 $\pm$ 50 cm^{-1}) but again there is some indication of the excitation of a much higher frequency mode (cf. 16·8 eV band of benzene spectrum). The excitation of the ring-breathing mode v_7 is suggested, but the frequency measured is perhaps too high since a rather larger reduction in frequency as compared with that in the molecular ground state was found for the equivalent vibration excited in the 16·8 eV ionization of benzene. If the vibration involved in each case were indeed the totally symmetric ring-breathing mode, however, this could be taken as evidence that the orbital involved is π throughout, in spite of other indications to the contrary. The larger value for the resonance integral β for benzene as compared with that for acetylene for example following from the assignment of the lowest π orbital I.P. as 16·8 eV could then possibly be associated with the characteristic 'aromatic' stability of benzene-like molecules which acetylene lacks. However, recent data on quantum defects for the Rydberg series leading to the 16·8 eV I.P. of a benzene definitely indicate a σ orbital.[21]

The effect of a chlorine atom in the α position of the furan ring is to decrease the first furan I.P. and increase the second as can be seen from a comparison of Figures 12.19 and 12.22. It seems probable that mesomeric interaction between the chlorine lone-pair orbitals and the highest-energy π orbital is responsible for the increased separation of the two lowest I.P.'s of chlorofuran as compared with furan. That the second I.P. is actually increased indicates that the inductive effect of the chlorine atom is somewhat stronger than the mesomeric effect. The appearance of the two chlorine (3p) bands exactly parallels the situation in chlorobenzene and the chloropyridines.

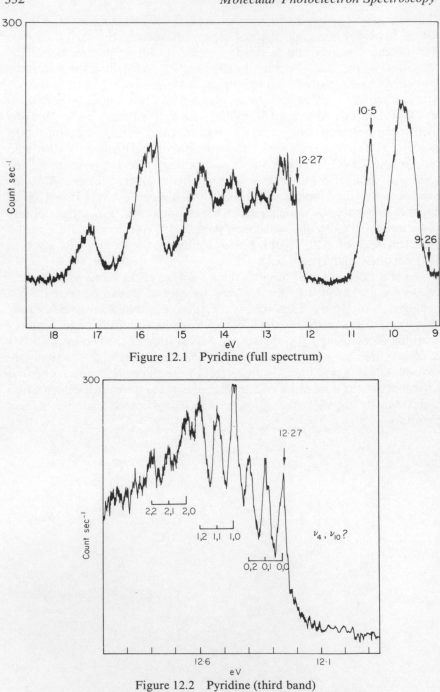

Figure 12.1   Pyridine (full spectrum)

Figure 12.2   Pyridine (third band)

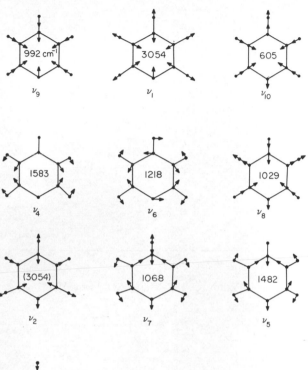

Figure 12.3 The normal modes of vibration of the pyridine molecule relevant to the interpretation of the photoelectron spectrum

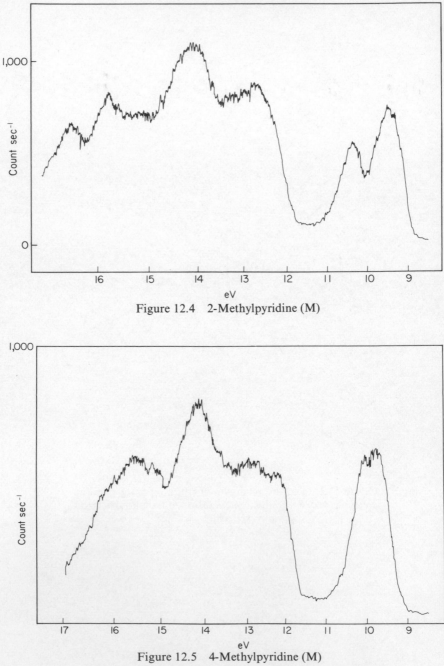

Figure 12.4   2-Methylpyridine (M)

Figure 12.5   4-Methylpyridine (M)

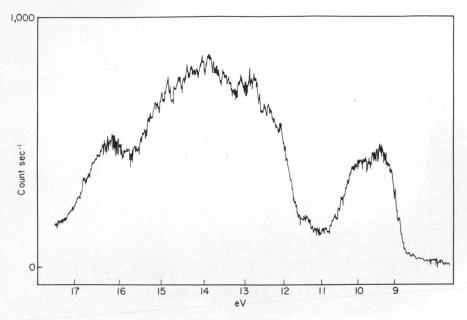

Figure 12.6 2,4-Dimethylpyridine (M)

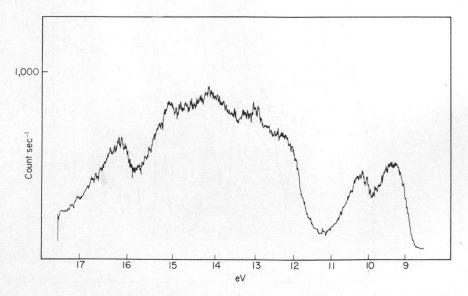

Figure 12.7 2,6-Dimethylpyridine (M)

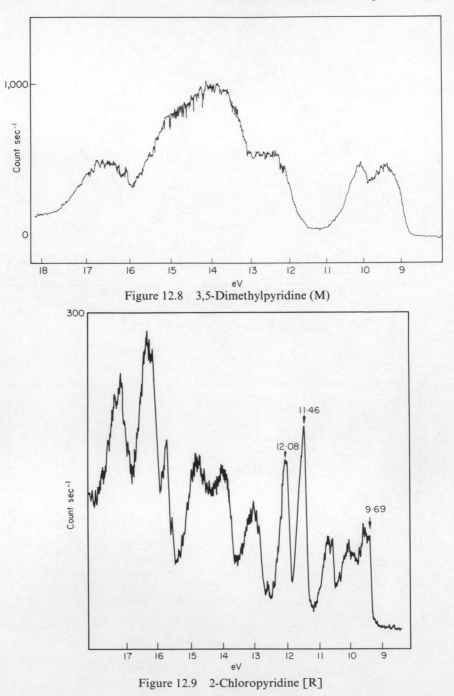

Figure 12.8   3,5-Dimethylpyridine (M)

Figure 12.9   2-Chloropyridine [R]

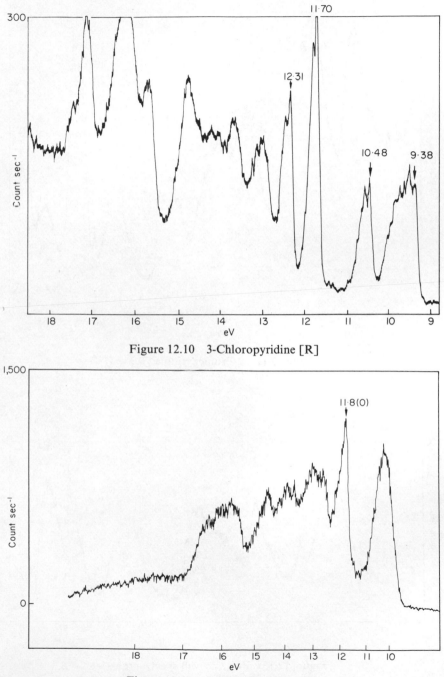

Figure 12.10   3-Chloropyridine [R]

Figure 12.11   4-Chloropyridine (M)

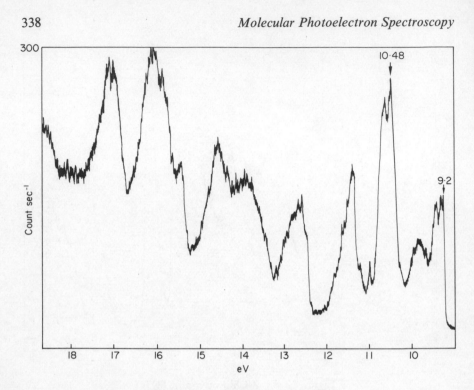

Figure 12.12   2-Bromopyridine [R]

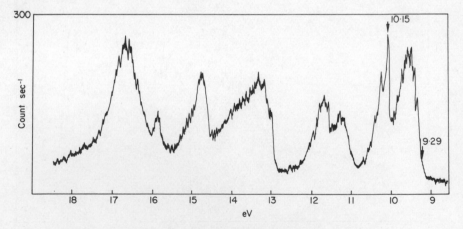

Figure 12.13   Pyrazine (full spectrum)

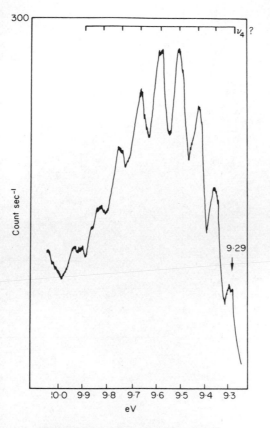

Figure 12.14   Pyrazine (first band)

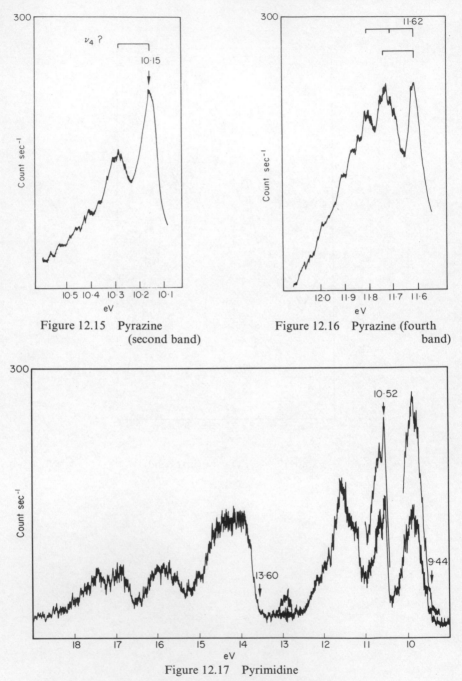

Figure 12.15   Pyrazine
(second band)

Figure 12.16   Pyrazine (fourth
band)

Figure 12.17   Pyrimidine

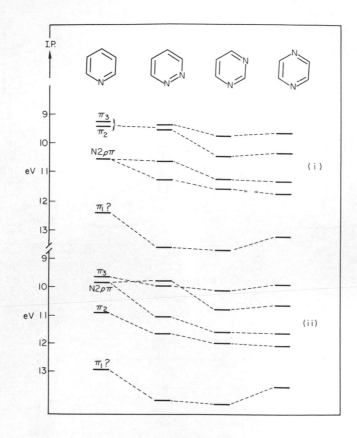

Figure 12.18 Correlation of the energy levels in pyridine and the benzene diazines inferred from their photoelectron spectra (see text)

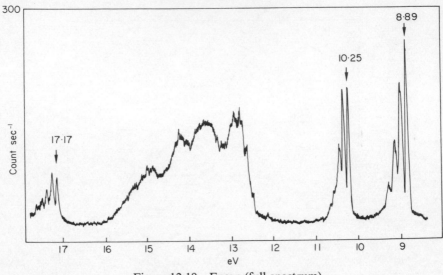

Figure 12.19   Furan (full spectrum)

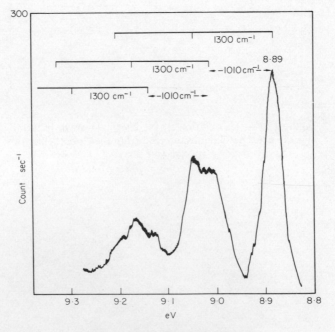

Figure 12.20   Furan (first band)

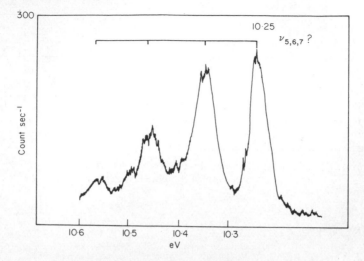

Figure 12.21   Furan (second band)

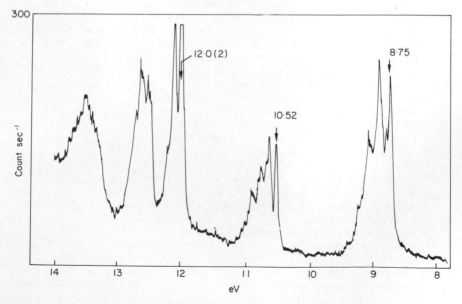

Figure 12.22   2-Chlorofuran

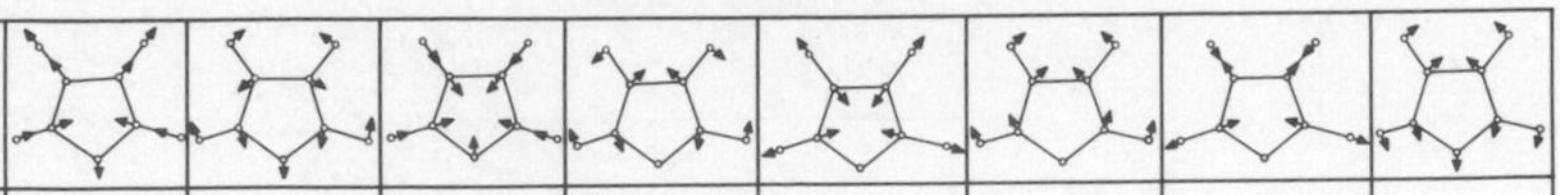

A	ν_1	ν_2	ν_3	ν_4	ν_5	ν_6	ν_7	ν_8
B	ν_8	ν_6	ν_7	ν_5	ν_4	ν_3	ν_2	ν_1
C	724	1067	994	1137	1381	1486	3090	3120

Figure 12.23   The normal modes of vibration of the pyridine molecule relevant to the interpretation of the photoelectron spectra

REFERENCES

1. A. J. Yencha and M. A. El-Sayed, *J. Chem. Phys.*, **48**, 3469 (1968).
2. E. Clementi, *J. Chem. Phys.*, **46**, 4731 (1967).
3. J. W. Emsley, *J. Chem. Soc., A*, **1968**, 1387.
4. J. A. Pople, D. P. Santry and G. A. Segal, *J. Chem. Phys. Suppl.*, **43**, 129 (1965); J. A. Pople and G. A. Segal, *J. Chem. Phys. Suppl.*, **43**, 136 (1965); J. A. Pople and G. A. Segal, *J. Chem. Phys.*, **44**, 3289 (1966).
5. M. D. Newton, F. P. Boer and W. N. Lipscomb, *J. Am. Chem. Soc.*, **88**, 2368 (1966).
6. J. del Bené and H. Jaffé, *J. Chem. Phys.*, **48**, 1807 (1968).
7. M. A. El-Sayed, M. Kasha and V. Tanaka, *J. Chem. Phys.*, **34**, 334 (1961).
8. I. Omura, H. Baba and K. Higasi, *J. Chem. Phys.*, **24**, 623 (1956); K. Higasi, I. Omura, H. Baba and I. Kanaoka, *Bull. Chem. Soc. Japan*, **30**, 633 (1957).
9. M. R. Basila and D. Clancy, *J. Phys. Chem.*, **67**, 1551 (1963).
10. V. G. Krishna and M. Chowdhury, *J. Phys. Chem.*, **67**, 1067 (1963).
11. M. I. Al-Joboury, *Ph.D. Thesis*, London University, 1964.
12. D. W. Turner, *Adv. Phys. Org. Chem.*, **4**, 31 (1966).
13. K. Watanabe, *J. Chem. Phys.*, **26**, 542 (1957).
14. D. W. Turner, *Tetrahedron Letters*, **35**, 3419 (1967).
15. E. Clementi, *J. Chem. Phys.*, **46**, 4737 (1967).
16. K. Watanabe and coworkers, *J. Chem. Phys.*, **29**, 48 (1958).
17. P. Clark, *Tetrahedron*, **24**, 3285 (1968).
18. C. Baker and D. W. Turner, *Proc. Roy. Soc.* (*London*), *Ser. A*, **308**, 19 (1968).
19. C. Baker and D. W. Turner, *Chem. Comm.*, **1967**, 797.
20. A. D. Baker, C. Baker, C. R. Brundle and D. W. Turner, *Intern. J. Mass Spectr. Ion. Phys.*, **1**, 285 (1968).
21. E. Lindholm, Private communication. (1968).
22. J. E. Parker and K. K. Innes, *J. Mol. Spectr*, **15**, 407 (1965).
23. H. W. Thompson and R. B. Temple, *Trans. Faraday Soc.*, **41**, 27 (1945).
24. G. Herzberg, 'Electronic Spectra and Molecular Structure of Polyatomic Molecules', *Molecular Spectra and Molecular Structure*, Vol. III, Van Nostrand, London and New York, 1966.
25. L. Corsin, B. J. Fox and R. C. Lord, *J. Chem. Phys.*, **21**, 1170 (1953).

CHAPTER 13

Hydrogen Cyanide and Related Compounds

1. HYDROGEN CYANIDE

This has been the object of many orbital energy calculations, the results of which are collected in Table 13.1. The 1π orbital is predicted in each case to be the highest-occupied molecular orbital and the first band in the photo-electron spectrum (Figures 13.1, 13.2), or part of it (see below), must arise from 1π ionization. The pattern of the vibrational peaks in the first band is complex, but a series in v_3, the C≡N stretching mode, can be picked out with frequency 1800 cm^{-1}, as compared with that in the molecular ground state of 2097 cm^{-1}.

Table 13.1   Orbital energies of hydrogen cyanide

Orbital	Calculated energies (eV)						Observed I.P.'s		
	Ref. 1	Ref. 2	Ref. 3	Ref. 4	Ref. 5	Ref. 6	Ref. 7	Ref. 8	Ref. 9
1π	-12.96	-13.8	-14.97	-13.74	-13.81	-13.11	13.60	13.91	13.59
5σ	-14.39	-15.2	-16.60	-15.18	-15.19	-14.48	~ 14		
4σ	-21.14	-21.7	-22.04	-21.66	-21.67	-22.05	19.06		
3σ	-33.14	-34.1	-35.10	-34.07	-34.07	-36.56			
2σ	-308.4		-312.9	-311.0	-311.0	-310.3			
1σ	-425.8		-428.8	-428.3	-428.3	-431.0			

The 20 eV band in the spectrum (Figure 13.1) is characteristic of those resulting from electron loss from a strongly bonding orbital. It is likely to be the 4σ orbital, which is strongly C—H bonding, and the vibrational frequency observed, 1690 cm^{-1}, can be assigned to the C—H stretching mode v_1 on the basis of the reduction to 1450 cm^{-1} in the corresponding band in the deuterium cyanide spectrum, $v_1 HCN^+/v_1 DCN^+$ being 1.18 as compared with $v_1 HCN/v_1 DCN$ of 1.25. The large reduction in the vibrational frequency on ionization is a consequence of the strongly bonding character of the 4σ orbital. In addition, the lack of discrete structure in the band above about 20.3 eV would be due to rapid dissociation of the molecular ion. The appearance potential of this fragment ion CN^+ can be calculated to be 19·4 eV, as is observed by photoionization.[9]

The 5σ orbital, which is essentially a nitrogen 'lone-pair', has been calculated to have an energy between 14·4 and 16·5 eV. However, the broad but very low 'hump' in the photoelectron spectrum (Figure 13.1) is unlikely to be assignable to this orbital, since the lone-pair ionization should give a band with simple vibrational fine structure. This however might not be the case if the molecular ion were unstable at about 16 eV when the fine structure would be expected to be broadened, in accordance with the Uncertainty Principle. But we know from mass spectrometry that HCN^+ is stable for at least 10^{-5} sec up to 19–20 eV. A band due to ionization from the 5σ level probably overlaps that due to ionization from the 1π level. The spectra of methyl cyanide and methyl isocyanide confirm this (see below) and it may account for the complexity of the peaks in the 13·6–14·5 eV region of the spectrum.

2. METHYL CYANIDE

Adiabatic I.P. 12·21 eV. When the hydrogen atom is replaced by a methyl group, the methyl 'pseudo π' orbitals can overlap with the $C\equiv N$ π orbitals, forming two new orbitals ('π_{CH_3}' + π_{CN}) and ('π_{CH_3}' $-$ π_{CN}). Thus the presence of the methyl group has a destabilizing effect on the $C\equiv N$ π orbitals and so increases the energy separation between the highest π and the nitrogen lone pair orbitals to such an extent that they become separately recognizable in the photoelectron spectrum (Figures 13.3–13.5).

Previous photoionization studies have given the first I.P. of methyl cyanide as 12·19 $\pm$ 0·01 eV,[9] 12·205 $\pm$ 0·004 eV[10] and 12·22 $\pm$ 0·01 eV,[11] and the orbital concerned as π. Three vibrational modes are excited in the ground state of the ion (Figure 13.4). They can be assigned to the $C\equiv N$ stretching mode (v_2), the symmetric CH_3 deformation (v_3) and the C—C stretching mode (v_4), with frequencies of 2010, 1430 and 810 cm^{-1}, respectively. Comparison of these frequencies with those occurring in the molecular ground state (2249, 1376 and 918 cm^{-1} respectively) indicates that the orbital is C—C and C—N bonding, but has some degree of C—H antibonding character. In the CD_3CN spectrum, the frequencies observed are 1990, 1070 and 810 cm^{-1}, giving rise to the ratios $v_2CH_3CN^+/v_2CD_3CN^+ = 1·01$, $v_3CH_3CN^+/v_3CD_3CN^+ = 1·34$ and $v_4CH_3CN^+/v_4CD_3CN^+ = 1·00$, being close to the theoretical values.

The sharp $0 \leftarrow 0$ transition at 13·14 eV can be assigned to electrons from the nitrogen 'lone-pair' σ orbital, and the vibrational mode excited with frequency 1290 cm^{-1} (970 cm^{-1} in CD_3CN, $v_H/v_D = 1·33$) is probably v_3.

Two modes are excited in the 15·5 eV band (Figure 13.5), with frequencies 1440 and 860 cm^{-1}, and are possibly the v_3 and v_4 modes, indicating that the orbital is C—H antibonding and C—C bonding. The corresponding band in the CD_3CN spectrum appears to contain only a simple series, but the individual peaks are about twice as broad as those in the CH_3CN band. The CD_3CN band supports the assignments of the vibrational modes in CH_3CN only in so far as expected peak separations corresponding to frequencies of 1080 cm^{-1} ($v_3CD_3CN^+ = v_3CH_3CN^+/1\cdot33$) and 810 cm^{-1} ($v_4CD_3CN^+ = v_4CH_3CN^+/1\cdot06$) can be accommodated by the series of broad peaks observed.

The vibrational modes excited in the various states of the ions CH_3CN^+ and CD_3CN^+, and their frequencies, are collected in Table 13.2.

Table 13.2   Vibrational modes excited in the ions CH_3CN^+ and CD_3CN^+

State of ion	Vibrational mode excited	Frequency (cm^{-1})		$\dfrac{v_{CH_3CN^+}}{v_{CD_3CN^+}}$
		CH_3CN^+	CD_3CN^+	
$\tilde{X}$	v_2 C≡N stretch	2010	1990	1·01
	v_3 symm. CH_3 def.	1430	1070	1·34
	v_4 C—C stretch	810	810	1·00
$\tilde{B}$	v_3 symm. CH_3 def.	1290	970	1·33
$\tilde{C}$	v_3 symm. CH_3 def.	1440	(1080)	1·33
	v_4 C—C stretch	860	(810)	1·06

The values in brackets were obtained by applying the theoretical reduction in frequency to those of CH_3CN^+.

3. METHYL ISOCYANIDE

The presence of a methyl–nitrogen bond in methyl isocyanide leads to a carbon atom localized lone-pair, at markedly higher energy than the nitrogen lone-pair of the cyanide. The lone-pair is thus destabilized to a greater extent than is the highest occupied π orbital. Therefore in the spectrum (Figures 13.6–13.8) the first band, adiabatic I.P. 11·27 eV, can be recognized as the carbon 'lone-pair' orbital. In fact the orbital has some C≡N antibonding character, since the C≡N stretching mode, v_2, is excited at a slightly higher frequency (2280 cm^{-1}) than occurs in the molecular ground-state (2161 cm^{-1}). In addition, the CH_3 deformation, v_3, is excited with a frequency (1410 cm^{-1}) equal to that which appears in the ground state of the molecule (1414 cm^{-1}).

The highest π band (adiabatic I.P. 12·24 eV), is slightly more bonding than in the cyanide, and a progression of peaks with spacing 1770 cm^{-1} (v_2) is resolved. The CH_3 deformation (v_3) is active to excitation in one quanta units, and appears with frequency 1130 cm^{-1}.

The band with $0 \leftarrow 0$ component at 15·59 eV consists of a simple vibrational series with peak separations of 750 cm^{-1}, which may be the C—N stretching mode v_4, which has a frequency of 928 cm^{-1} in the molecule.

4. CYANOGEN

The electronic structure of cyanogen has been calculated by Clementi[4] who found that the energy of the highest occupied σ orbital ($5\sigma_g$) was intermediate in value between those of the two occupied π orbitals. The photoelectron spectrum[7] (Figures 13.9–13.11) shows clearly that the ionization energies of the *two* highest σ orbitals ($4\sigma_u$) and ($5\sigma_g$), come between those of the two π orbitals. The two σ orbitals are the out-of-phase and in-phase combinations of the nitrogen nonbonding orbitals, which give rise to very intense $0 \leftarrow 0$ transitions upon ionization, at energies 14·49 and 14·86 eV. The first of these has associated with it a vibrational progression in v_1, the symmetric C≡N stretching mode, excited with a frequency 1860 cm^{-1}, as compared with the ground state frequency of 2336 cm^{-1}.

The first and fourth bands in the spectrum, at adiabatic I.P.'s 13·36 and 15·47 eV can, by comparison of the spectra of the isoelectronic molecules diacetylene and cyanoacetylene (Chapter 6), be readily recognized as referring to electrons from the $1\pi_g$ and $1\pi_u$ orbitals, respectively. In the ($^2\Pi_g$) ground state of the ion the v_1 vibrational mode is excited with a frequency of 2120 cm^{-1}. A frequency of 2178 cm^{-1} has been observed in photoionization efficiency curves by Dibeler.[12]

As with diacetylene and cyanoacetylene, the $^2\Pi_u$ state of the ion has vibrational modes v_1 and v_2 excited. The expected C≡N stretching mode v_1 has a frequency of 2020 cm^{-1} and the C—C stretching mode v_2 has a frequency of 710 cm^{-1}. In addition another mode appears to have a frequency of 570 cm^{-1} which seems only comparable with the symmetric bending mode v_4. This has been observed weakly excited in the 3020–2400 Å region of the absorption spectrum,[13] and the photoelectron spectrum shows that the mode in question appears in units of one quantum. This points to the possibility that the $^2\Pi_u$ state of $C_2N_2^+$ is *trans* bent with only a small potential barrier in the linear form.

The correlation between the ionization energies of cyanogen and linear unsaturated hydrocarbons and some of their aza derivatives is illustrated in Chapter 6, and is discussed more fully by Baker and Turner in Reference 7.

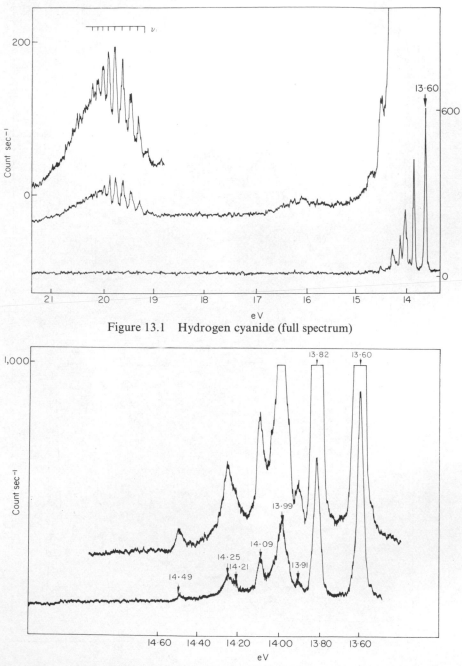

Figure 13.1   Hydrogen cyanide (full spectrum)

Figure 13.2   Hydrogen cyanide (first band)

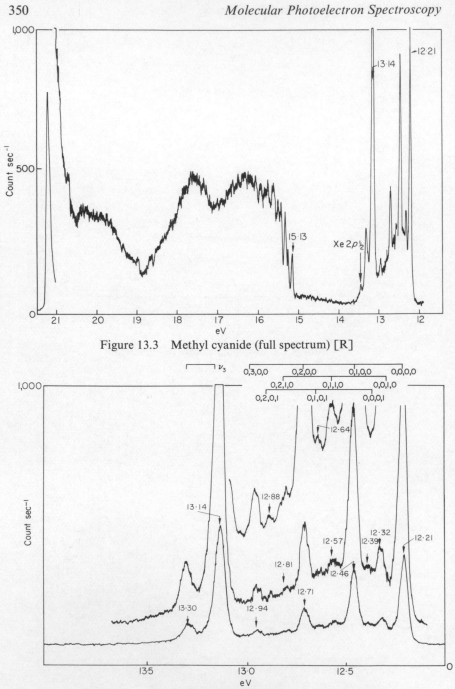

Figure 13.3   Methyl cyanide (full spectrum) [R]

Figure 13.4   Methyl cyanide (first and second bands) [R]

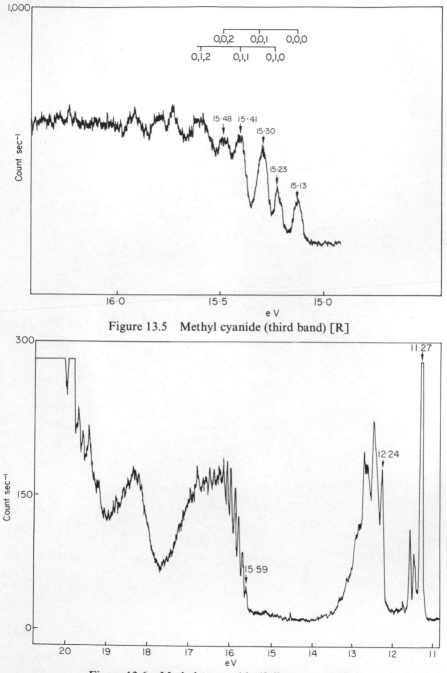

Figure 13.5 Methyl cyanide (third band) [R]

Figure 13.6 Methyl *iso*cyanide (full spectrum) [R]

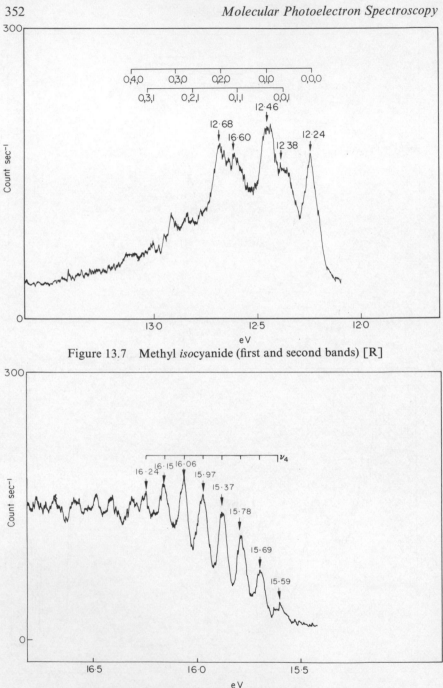

Figure 13.7 Methyl *iso*cyanide (first and second bands) [R]

Figure 13.8 Methyl *iso*cyanide (full spectrum) [R]

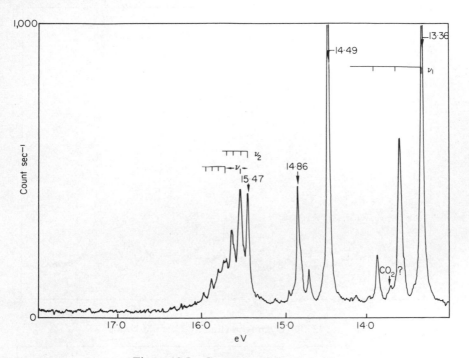

Figure 13.9   Cyanogen (full spectrum)

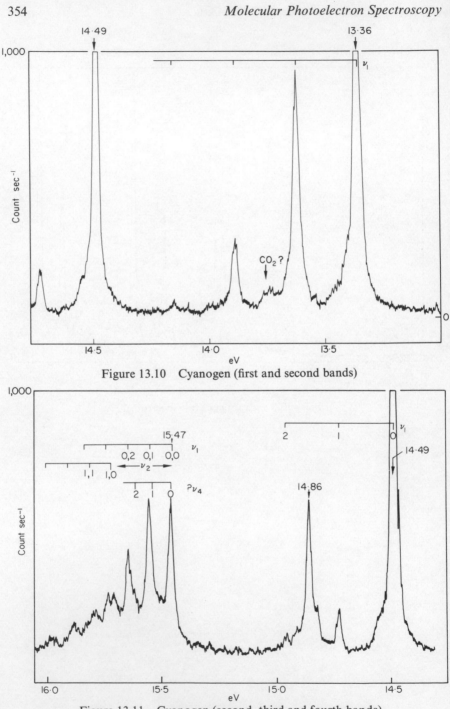

Figure 13.10   Cyanogen (first and second bands)

Figure 13.11   Cyanogen (second, third and fourth bands)

REFERENCES

1. W. E. Palke and W. N. Lipscomb, *J. Am. Chem. Soc.*, **88**, 2384 (1966).
2. D. Peters, *J. Chem. Phys.*, **45**, 3474 (1966).
3. D. C. Pan and L. C. Allen, *J. Chem. Phys.*, **46**, 1797 (1967).
4. E. Clementi and H. Clementi, *J. Chem. Phys.*, **36**, 2824 (1962).
5. A. D. McLean, *J. Chem. Phys.*, **37**, 627 (1962).
6. J. B. Moffat, *Chem. Comm.*, **7**, 89 (1966).
7. C. Baker and D. W. Turner, *Proc. Roy. Soc. (London), Ser. A*, **308**, 19 (1968).
8. J. D. Morrison and A. J. C. Nicholson, *J. Chem. Phys.*, **20**, 1021 (1952).
9. V. H. Dibeler and S. K. Liston, *J. Chem. Phys.*, **48**, 4765 (1968).
10. A. J. C. Nicholson, *J. Chem. Phys.*, **43**, 1171 (1965).
11. K. Watanabe, T. Nakayana and J. Mottl, *J. Quant. Spectr. Radiative Transfer*, **2**, 369 (1962).
12. V. H. Dibeler and S. K. Liston, *J. Chem. Phys.*, **47**, 4548 (1967).
13. S. C. Woo and T. K. Liu, *J. Chem. Phys.*, **5**, 161 (1937).

CHAPTER 14

Miscellaneous Inorganic Compounds

1. AMMONIA AND PHOSPHINE

(A) Ammonia

The photoelectron spectrum of NH_3 has been published on two previous occasions. The original publication,[1] in 1964, concerned a low resolution cylindrical-grid spectrum in which no vibrational fine structure could be resolved. The second publication reported a higher resolution study by Brehm and Puttkamer[2] in which the photoelectrons were analysed by a linear retarding-field system. Their apparatus also incorporated a method of measuring the electron energy spectrum in coincidence with the mass-analysed photoions, though under these conditions a much lower electron energy resolution was accepted.

In the present spectrum (Figures 14.1, 14.2), the vibrational structure on the first band is fully resolved, and a certain amount of structure on the second band is also revealed. The experimental I.P.'s and vibrational frequencies are recorded in Table 14.1, together with values for the ground

Table 14.1 Ionization potentials and observed vibrational frequencies
for ammonia

Photo-electron band	Electronic state	Adiabatic I.P. (eV)	Observed vibrational frequency (cm^{-1})			
			$1^{(a)}$	$2^{(b)}$	3	4
—	Ground molecular state[4] $\ldots(1e)^4\,(3a_1)^2$	—	3337	$968^{(c)}$ $933^{(c)}$	3444	1627
1st	Ground ionic state $\ldots(1e')^4\,(1a_2'')^1$	10·16(0)	—	900	—	—
2nd	1st excited ionic state $\ldots(1e')^3\,(1a_2'')^2$	14·8(7)	1800?	—	—	—

(a) N—H stretching mode.   (b) N—H out of plane bending mode.
(c) These are the two inversion components.[4]

state of the molecule. Experimental Franck–Condon factors for the first band have also been measured, and are recorded in Table 14.2 with the calculated values due to Botter and Rosenstock[3] and the experimental values obtained by Brehm and Puttkamer.[2]

Table 14.2 Experimental and calculated Franck–Condon factors for transitions to the ground ionic state of NH_3

[2]Experimental	Experimental F–C factors		Calculated values[3] $(\theta = 120°\ r_e = 0.934Å^{(a)})$
	Present work	Ref. 2	
0	—	0·01	0·016
1	0·14	0·06	0·109
2	0·37	0·17	0·335
3	0·65	0·34	0·662
4	0·86	0·55	0·995
5	0·94	0·76	0·100
6	1·00	1·00	0·833
7	0·94	—	0·556
8	0·65	—	—
9	0·35	—	—
10	0·27	—	—
11	0·14	—	—
12	0·08	—	—
13	0·06	—	—
14	0·03	—	—

The ground state configuration of the NH_3 molecule may be represented as[4]:

$$...(2a_1)^2\ (1e)^4\ (3a_1)^2$$

where the $(3a_1)$ orbital is largely nonbonding on the nitrogen atom, $(2p_zN)$, and the doubly degenerate $(1e)$ orbital is strongly N—H bonding.

Several calculations[5–10] on the MO structure of NH_3 have been carried out and the energy levels are indicated in Table 14.3, where they are compared with the experimental vertical ionization potentials.

Table 14.3 Orbital energy levels of NH_3

Orbital	Calculated energy level (eV)				Experimental vertical I.P. (eV)
	Ref. 5	Ref. 6	Ref. 8	Ref. 10	
$(3a_1)$	−14·0	− 9·94	−11·3	−10·26	10·85
$(1e)$	−19·3	−16·3	−16·3	−15·5	15·8
$(2a_1)$	−32·7	−31·2	−33·0	−30·0	21·22

The two bands in the photoelectron spectrum will now be considered in some detail.

The first band (Figure 14.2 Adiabatic I.P. 10·16(0) eV Vertical I.P. 10·85 eV) consists of a long single series in the out-of-plane bending vibrational mode, v_2, having at least 15 member peaks. This is not the type of band which would at first sight be expected for the removal of a supposedly 'nonbonding' electron. Walsh and Warsop[11] showed that the electronically excited states of NH_3 and in the limit the ground state of NH_3^+ were planar. Since this involves a large change in the bond angle ($107° \rightarrow 120°$) on going from the pyrimidal ground molecular state, one would expect a long series in the bending vibration,[11] just as is observed in the photoelectron spectrum and also in the Rydberg bands in absorption.[11] Brehm and Puttkamer[2] observed the first seven vibrational steps in their photoelectron spectrum, and the photoionization efficiency curve of Dibeler and coworkers[12] exhibits nine steps in the region of the first I.P., though there is probably some contribution from autoionization also.

According to the selection rules,[9] since the v_2 vibration is non-totally symmetric with respect to the D_{3h} symmetry of the ion, transitions involving excitation of v_2 are only allowed when the upper and lower rovibronic states have the same symmetry. Normally this would mean that only 0, 2, 4, etc., quanta of v_2 would be allowed in the ionic state. In this case however, all values of v_2 are allowed because transitions may take place from one or other component of the inversion-doubled ground electronic state.[4] The fact that the anharmonicity constant of the series is negative ($v_{13}'-v_{14}'$ vibrational interval is approximately 1050 cm^{-1}) probably indicates a fairly large quartic term in the potential function of the ion.

The first clearly identifiable peak of the series falls at 10·27(6) eV. Though this value is lower than that obtained from electron impact work,[13-15] 10·4–10·6 eV, it is higher than the values quoted from photoionization,[12] Rydberg series limits,[16] and the two previously published photoelectron spectra.[1,2] These are all in the region of 10·16 eV. This is surprising, as all previous experience has shown that it is easier to detect the true adiabatic I.P.'s by high-resolution photoelectron spectroscopy when the first few vibrational levels are only weakly populated than by other techniques.

Botter and Rosenstock[3] have shown from their calculations of Franck–Condon factors that the value of the $0 \leftarrow 0$ transition probability, (Table 14.3), is always so low (assuming a planar structure and including r_e as a variable parameter in the calculation) that it would be difficult to detect by photoionization experiments. We consider therefore that the first clearly observed member of the series in our spectrum relates to the $1 \leftarrow 0$ (v_2) transition, and that the $0 \leftarrow 0$ transition lies at 10·16(0) eV. It can be seen from Table 14.2 that the remaining experimental Franck–Condon factors, though only roughly measured, disagree badly with the calculated values and Brehm and Puttkamer's values.[3] A more careful study of the experimental Franck–Condon factors and a definite establishment of the position of the $0 \leftarrow 0$ transition should be useful in determining the parameters of the ion by comparison with calculated values for a varying set of parameters.

In the second band (Adiabatic I.P. 14·8(7) eV, Vertical I.P. 15·8 eV) the exact position of the $0 \leftarrow 0$ transition is again in doubt since there is a slight possibility that one or two peaks of very low intensity may not have been detected.

14·8(7) eV is a lower value than that obtained from the less accurate previous photoelectron determinations,[1,2] 15·02 eV, and than the values claimed from breaks in electron impact[13] and photon impact[17] efficiency curves, 15·31 eV and 15·01 eV. The most recent and probably the most reliable photon impact work[12] shows no sign of a second threshold for NH_3^+ formation in this region, though it may be masked by the unresolved peaks due to autoionizing levels.

The vibrational series to the low I.P. side of the band consists of four or five very diffuse 'steps', which are detectable only up to approximately 15·8 eV, a value which is very close to the appearance potential of NH_2^+, 15·73 eV.[12] It is possible that the vibration excited is the N—H stretching mode with a large reduction in frequency compared to that of the ground state of the molecule. The broadness of the vibrational peaks, the breaking off of the series at the A.P. of NH_2^+, together with the sharp decrease in molecular ion intensity at the same energy, and in particular the work of Brehm and Puttkamer[2] which showed that all ejected photoelectrons above an I.P. of 15·7 eV were in coincidence with the NH_2^+ fragment, all suggest that NH_2^+ is being formed by predissociation by curve crossing (cf. OH^+ from H_2O^+, and fragment ions from the linear triatomic molecules).

As was found for the linear triatomic molecules,[18] the A.P. of the fragment falls close to the accepted thermochemical limit for the process:

$$NH_3 \xrightarrow{h\nu} NH_2^+ + H + e$$
$$AP\ NH_2^+ = \text{I.P.}\ (NH_2) + D(NH_2 - H)$$
$$= 11\cdot22\ eV^{12} + 4\cdot52\ eV^{12}$$
$$= 15\cdot74\ eV$$

The second band in the photoelectron spectrum has a noticeably flattened 'summit', which may indicate Jahn–Teller splitting of this ionic state (cf. Methane, page 164). This splitting could amount to approximately 0·5 eV.

(B) Phosphine

The photoelectron spectrum of PH_3 (Figure 14.3) has not been published previously, and it is included here for the sake of comparison with NH_3. The spectra are indeed very similar, as expected from their basically similar molecular orbital structures, each consisting of two bands showing vibrational fine structure. As can be seen from Table 14.4, the calculated energy levels of the highest filled orbitals for PH_3 are in remarkably good agreement with the experimental vertical ionization potentials

Table 14.4   Orbital energy levels of PH_3

Orbital	Calculated energy level (eV) Ref. 8	Experimental vertical I.P. (eV)
$(5a_1)$	$-10\cdot0$	$9\cdot9$
$(2e)$	$-13\cdot1$	$13\cdot0$
$(4a_1)$	$-22\cdot8$	$21\cdot2$

The first band (Figure 14.4, Adiabatic I.P. $9\cdot27$ eV, Vertical I.P. $9\cdot9$ eV) exhibits a long vibrational series in v_2, as with NH_3, but in this case there is a very large reduction in frequency compared to that of the ground molecular state (420 cm$^{-1} \leftarrow 992$ cm^{-1}). A similar reduction has been observed in the frequencies contained in Rydberg levels detected in absorption. If one assumes a planar structure for PH_3^+ in the ground ionic state, then the greater reduction in frequency of v_2, compared to that of NH_3^+, may be due to the greater angle change that has occurred ($120° \leftarrow 93°$), cf. ($120° \leftarrow 107°$). The structure is not sufficiently well-resolved, nor was the experimental count-rate high enough to enable experimental Franck–Condon factors to be measured with accuracy. The experimental I.P. is considerably lower than that reported from electron impact work,[19] $10\cdot2$ eV, as might be expected.

The second band (Figure 14.5 Adiabatic I.P. $12\cdot19$ eV, Vertical I.P. $13\cdot0$ eV) is very similar to that in NH_3^+, showing 4 or 5 broad vibrational components of a series, which then breaks off near the vertical I.P. The vibrational spacing, 1600 cm^{-1}, probably represents the P—H stretching frequency (2323 cm^{-1} ground state molecule[4]). The energy at which the vibrational series breaks off very nearly coincides with the appearance potential of the fragment ion, PH_2^+, $13\cdot2$ eV, as measured by electron impact,[19] and the mechanism of formation is probably very similar to that of NH_2^+ from NH_3.

2. CARBON SUBOXIDE

Previous work on the ionization potential of carbon suboxide has included the photoionization experiments of Roebber,[20] who obtained a first I.P. of

10·60 eV, in agreement with the Rydberg series value.[20, 21] No information regarding the inner I.P.'s was obtainable using these techniques.

The electronic structure of carbon suboxide has been reported.[22, 23] The twelve π electrons are distributed into three doubly degenerate orbitals, $(1\pi_u)$, $(1\pi_g)$ and $(2\pi_u)$, and the photoelectron spectrum (Figure 14.6) indicates that the orbital of highest energy, almost certainly $(2\pi_u)$, is rather nonbonding. The vibrational fine structure associated with the first spectral band, adiabatic I.P. 10·60 eV (Figure 14.7) can be analysed in terms of excitation of the symmetric C=O stretching mode v_1, with a frequency of 1950 cm^{-1}, and the symmetric C=C stretching mode v_2, with a frequency of 660 cm^{-1}. These frequencies may be compared with those occurring in the ground state of the molecule, viz. 2200 and 830 cm^{-1} respectively.[24] Similar frequencies have been reported[21] in the Rydberg levels leading to the first I.P.

The vibrational structure associated with the second and third bands (Figure 14.8) is extremely diffuse, but it is possible to pick out a series of peaks in each, the corresponding vibrational frequencies (v_2) being 700 cm^{-1} in the first excited ionic state, and 650 cm^{-1} in the second excited state. The breadth of the few vibrational peaks, and the general lack of well resolved fine structure could be due to the excitation of low frequency bending modes[25] or to the result of the crossing of the potential energy surfaces of the first and second excited states of the ion with those leading to the fragment ions C_2O^+ and C_3O^+, which have appearance potentials at 15·1 and 15·9 eV respectively.[26]

The fourth band (Figure 14.9) admits of no simple analysis. The two strong peaks at 16·98 and 17·25 eV could be the first two members of a progression in v_1, but the higher members of appropriate intensity are missing. If their absence is not due to the sudden onset of dissociative broadening, the two intense peaks must be assigned to electrons from weakly bonding orbitals, and the shoulder near 17·6 eV, to a more strongly bonding orbital.

The relative order of the two lower π orbitals and the highest occupied σ orbitals cannot be deduced from the spectrum. Hückel MO calculations[21] place the energies of $(1\pi_u)$ and $(1\pi_g)$ orbitals at $-15\cdot2$ and $-14\cdot8$ eV.

3. METAL CARBONYLS

When a transition-metal atom is surrounded by an octahedral array of identical ligands, the ligand field theory tells us that the metal d-orbital energies will split into two groups, those of the d_{xy} and d_{xy} orbitals being reduced $(2t_{2g})$, and those of the $d_{x^2-y^2}$ and d_{z^2} being increased $(3e_g)$. In the spectra of chromium, molybdenum and tungsten hexacarbonyls (Figures 14.10, 14.11 and 14.12) the band at the ionization potential of about 8·4 eV can be assigned to the triply degenerate t_{2g} orbital. Electron impact[27] and photoionization[28] experiments have placed the order of first I.P.'s as $Cr <$ $Mo < W$. This is not evident from the photoelectron spectra which suggest them to be identical.

There is obviously (Figures 14.10–14.12) a large difference in energy between the t_{2g} orbital and the second highest occupied orbital, which may be the $2e_g$ metal orbital or the $3t_{1u}$ ligand orbitals. Calculations[29] of the orbital energies suggest that it is the $2e_g$ orbital.

The energies of the ligand orbitals seem to be very similar to those observed in the free carbon monoxide molecule, though their bonding characteristics are altered somewhat. Thus the oxygen $2p\sigma$ orbital I.P.'s fall in the 14–16 eV band and the $C{=}O$ π orbitals may be assigned to the band at about 17·6 eV. The broad band at about 19·4 eV may correspond to the 19·6 eV band in the carbon monoxide spectrum (Figure 3.10).

If one of the ligands is changed, this may have the effect of raising the degeneracy of the $2t_{2g}$ orbital, and this in fact occurs in $Mn(CO)_5CF_3$. The metal d_{xz} and d_{yz} orbitals remain degenerate, but the energy of the d_{xy} is higher. (The CF_3 ligand is considered as approaching along the z-axis.) Two bands at low I.P. are therefore observed in the photoelectron spectrum (Figure 14.13), at 9·4 and 10·4 eV, with areas in the ratio of approximately 2:1 as expected from the remaining degeneracy ($d_{xz} = d_{yz} \neq d_{xy}$).

4. METALLOCENES

The biscyclopentadienyl compounds of the transition metals of which the best known member is ferrocene are described as having a sandwich structure. The metal atom occupies the central position between the planes of two opposed cyclopentadienyl residues. As we have already remarked above in the case of the carbonyls the degeneracy of the $3d$ metal atomic orbitals is sensitive to the symmetry of the electrostatic field which the ligands produce. In the symmetry of the ferrocene molecule (D_{5d}) the five $3d$ orbitals of iron are split into groups. The middle group of these form the highest occupied orbitals in ferrocene, e_{2g}, a doubly degenerate pair and a_{1g}, nondegenerate. This splitting pattern is reflected in the highest energy band of the spectrum (Figure 14.15) by the approximately 2:1 ratio of the electron flux in the two components.

In the analogous cobalt and nickel compounds (Figures 14.16, 14.17), one and two additional electrons have been allotted to the higher hitherto unoccupied orbital and we expect additional bands in the photoelectron spectrum at lower ionization potential in a manner reminiscent of nitric oxide and oxygen when compared with nitrogen. This analogy is useful in that we are reminded that in proceeding from nitrogen to these two molecules we move from a closed shell, singlet molecule to open shell structures of nonzero multiplicity. In these cases we found that each band, other than the first, was a multiplet. The same multiplicity is apparent in the photoelectron spectra of cobaltocene and nickelocene since these also have one and two electrons respectively outside the ferrocene 'closed shell'. It also appears in chromocene, which has one less (Figure 14.14).

It has been pointed out by Green[20] that the chemistry of cobaltocene is easily understood in terms of this isolated outer electron which confers alkali metal-like properties. It is also to be noted that cobaltocene has significantly the lowest ionization potential in this group and indeed the lowest molecular ionization potential yet measured.

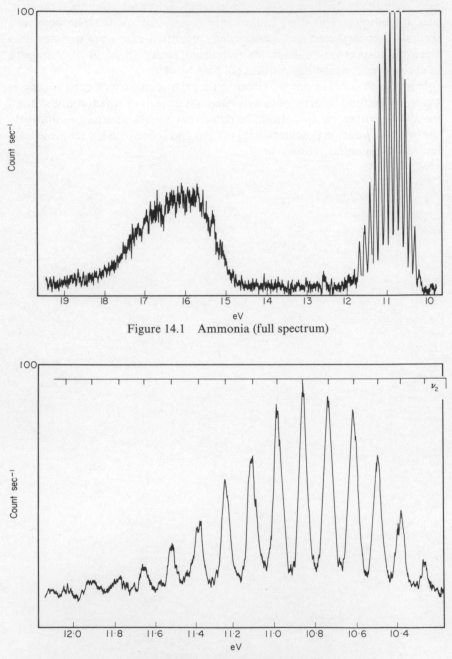

Figure 14.1   Ammonia (full spectrum)

Figure 14.2   Ammonia (first band)

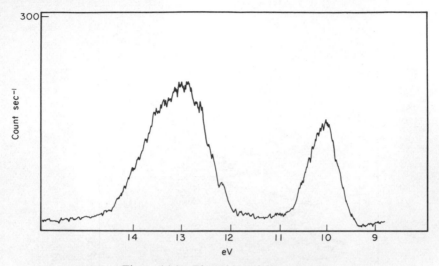
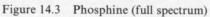

Figure 14.3   Phosphine (full spectrum)

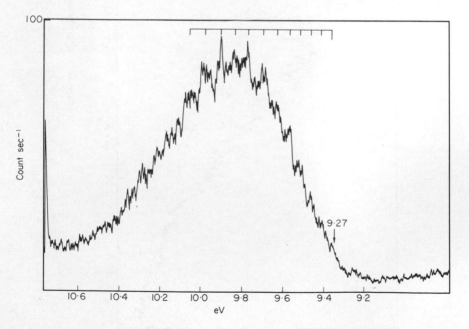
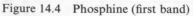

Figure 14.4   Phosphine (first band)

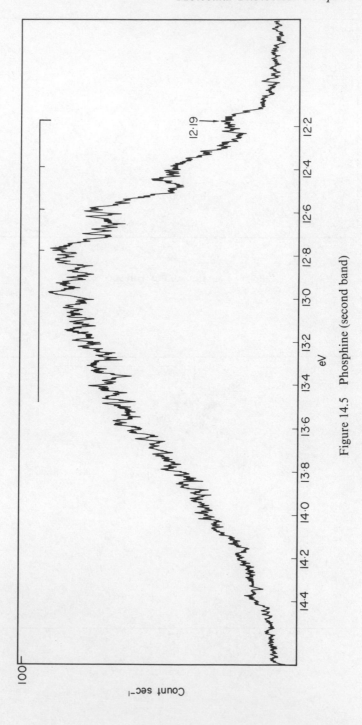

Count sec⁻¹

100

eV

12·19

Figure 14.5 Phosphine (second band)

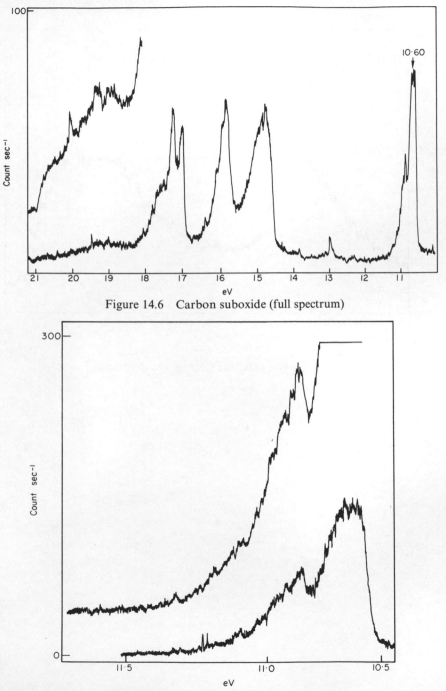

Figure 14.6   Carbon suboxide (full spectrum)

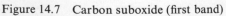

Figure 14.7   Carbon suboxide (first band)

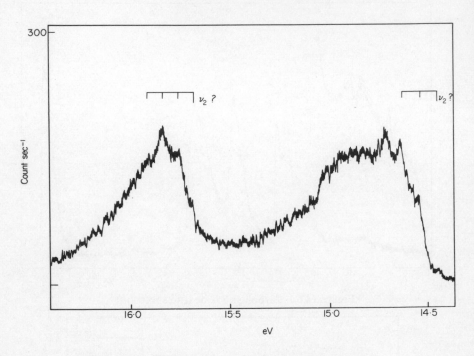

Figure 14.8   Carbon suboxide (second and third bands)

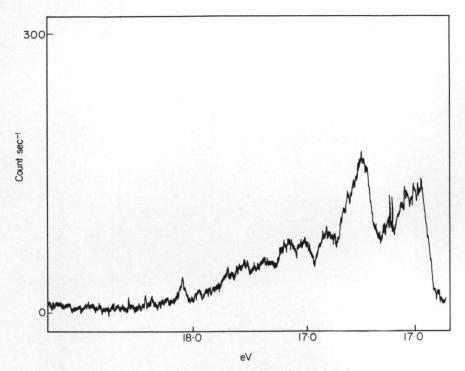

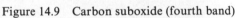

Figure 14.9 Carbon suboxide (fourth band)

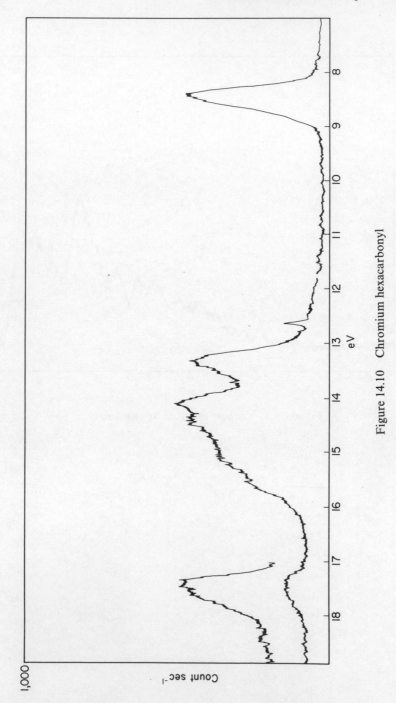

Figure 14.10   Chromium hexacarbonyl

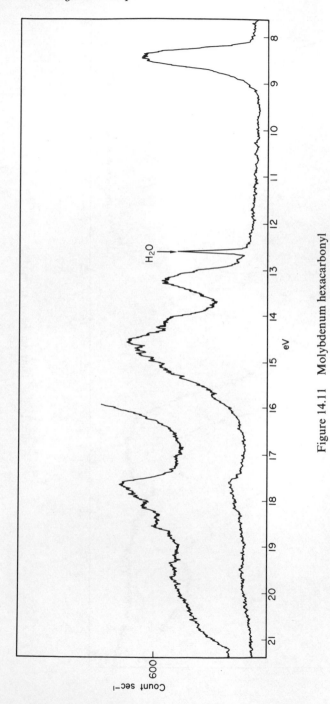

Figure 14.11 Molybdenum hexacarbonyl

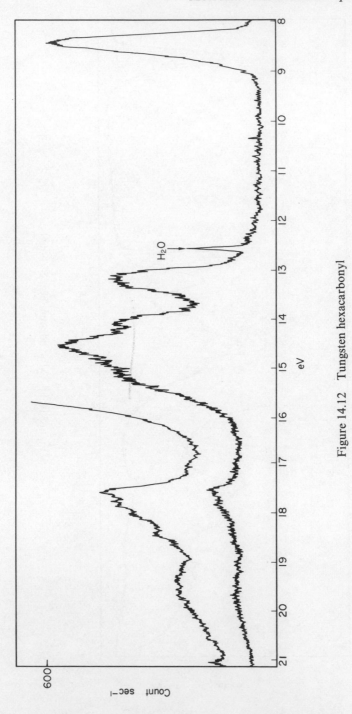

Figure 14.12   Tungsten hexacarbonyl

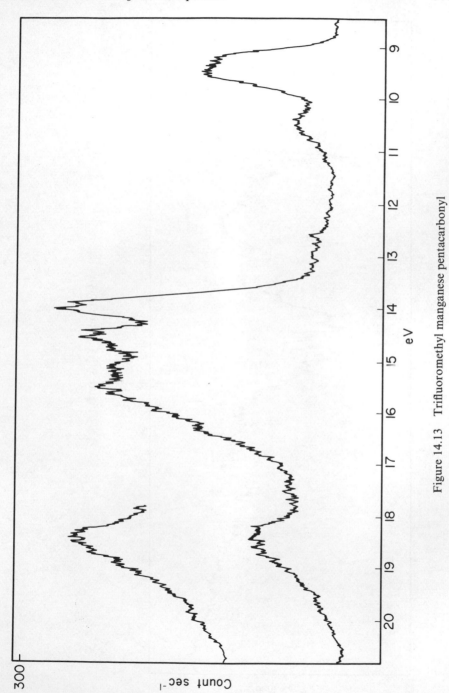

Figure 14.13 Trifluoromethyl manganese pentacarbonyl

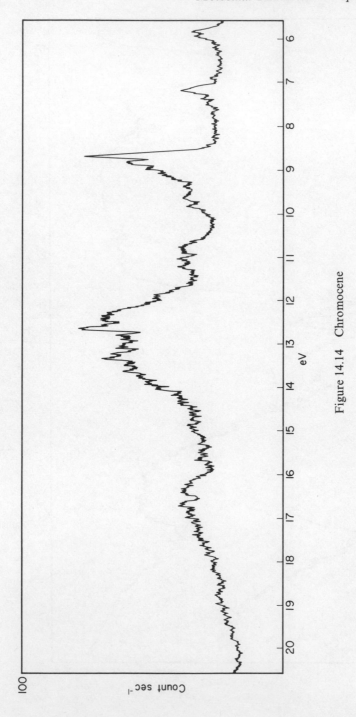

Figure 14.14 Chromocene

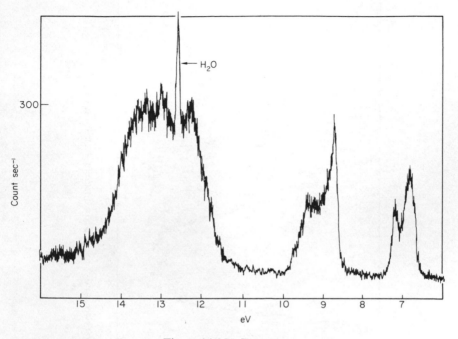

Figure 14.15   Ferrocene

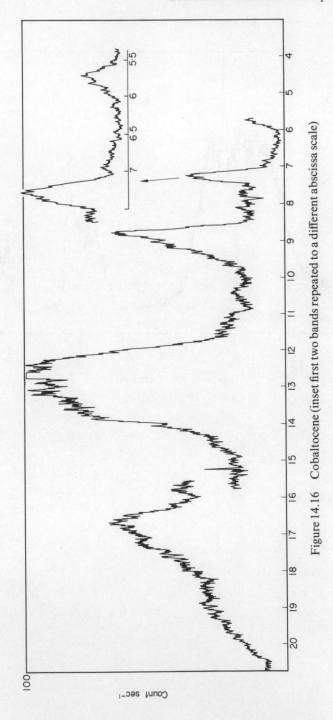

Figure 14.16   Cobaltocene (inset first two bands repeated to a different abscissa scale)

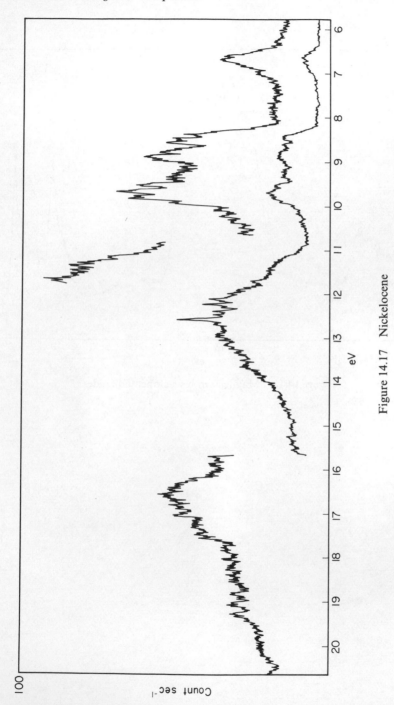

Figure 14.17 Nickelocene

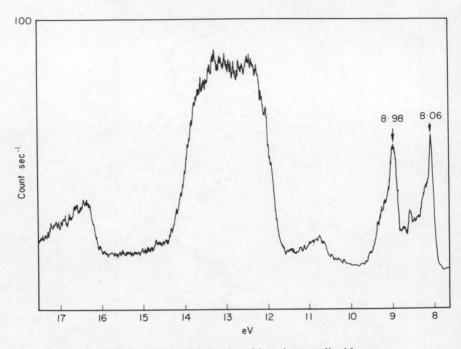

Figure 14.18   Magnesium biscyclopentadienide

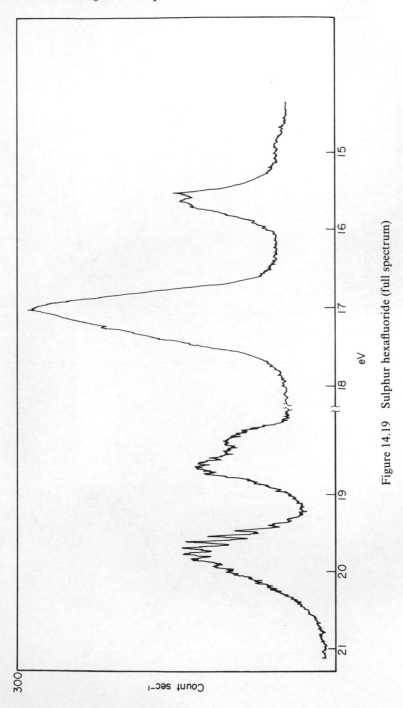

Figure 14.19   Sulphur hexafluoride (full spectrum)

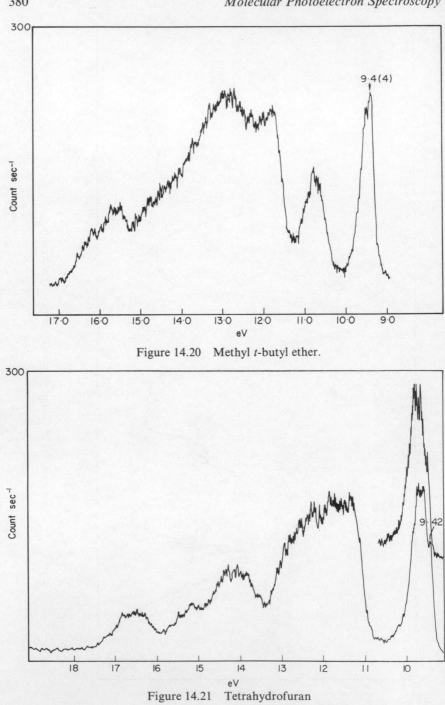

Figure 14.20   Methyl *t*-butyl ether.

Figure 14.21   Tetrahydrofuran

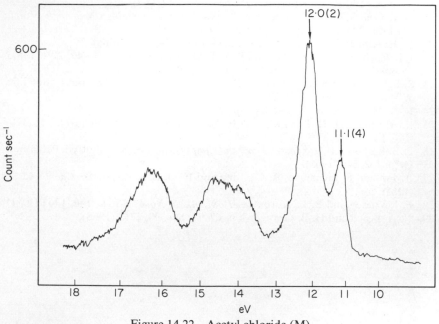

Figure 14.22   Acetyl chloride (M)

REFERENCES

1. M. I. Al-Joboury and D. W. Turner, *J. Chem. Soc.*, **1964**, 4434.
2. B. Brehm and E. Von Puttkamer, *Intern. Conf. Mass Spectr.*, Berlin, Sept. 1967.
3. R. Botter and H. M. Rosenstock, *Intern. Conf. Mass Spectr.*, Berlin, Sept. 1967.
4. G. Herzberg, *Molecular Spectra and Molecular Structure*, Vol. III, Van Nostrand, New York, 1966, pp. 390, 515.
5. H. Kaplan, *J. Chem. Phys.*, **26**, 1704 (1957).
6. A. B. F. Duncan, *J. Chem. Phys.*, **27**, 423 (1957).
7. D. Peters, *J. Chem. Phys.*, **36**, 2743 (1962).
8. R. Moccia, *J. Chem. Phys.*, **40**, 2176 (1964).
9. M. Krauss, *J. Res. Nat. Bur. Std.*, *A*, **68**, 635 (1964).
10. U. Kaldor and I. Shavitt, *J. Chem. Phys.*, **45**, 888 (1966).
11. A. D. Walsh and P. A. Worsop, *Trans. Faraday Soc.*, **57**, 345 (1961).
12. V. H. Dibeler, A. Walker and H. M. Rosenstock, *J. Res. Nat. Bur. Std.*, *A*, **70**, 459 (1966).
13. D. C. Frost and C. A. McDowell, *Can. J. Chem.*, **36**, 39 (1958).
14. M. M. Mann, A. Hustrulid and J. T. Tate, *Phys. Rev.*, **58**, 340 (1940).
15. J. D. Morrison and A. J. C. Nichollson, *J. Chem. Phys.*, **20**, 1021 (1952).
16. K. Watanabe and S. P. Sood, *Sci. Light*, (*Tokyo*), **14**, 36 (1965).
17. W. C. Walker and G. L. Weissler, *J. Chem. Phys.*, **23**, 1540 (1955).
18. V. H. Dibeler and J. A. Walker, *Intern. Conf. Mass Spectr.*, Berlin, Sept. 1967.

19. Y. Wado and R. W. Kiser, *Inorg. Chem.*, **3**, 174 (1964).
20. G. E. Coates, M. L. H. Green and K. Wade, *Organometallic Compounds*, Vol. 2, 3rd Edn., Methuen, London, 1968, p. 105.
   H. H. Kim and J. L. Roebber, *J. Chem. Phys.*, **44**, 1709 (1966).
21. J. L. Roebber, J. C. Larrabee and R. E. Huffman, *J. Chem. Phys.*, **46**, 4594 (1967).
22. C. Baker and D. W. Turner, *Chem. Comm.*, 400 (1968).
23. S. Bell, T. S. Varadarajan, A. D. Walsh, P. A. Warsop, J. Lee and L. Sutcliffe, *J. Mol. Spectr.*, **21**, 42 (1966).
24. F. A. Miller and W. G. Fateley, *Spectrochim. Acta*, **20**, 253 (1964).
25. F. A. Miller, D. H. Lemmon and R. E. Witowski, *Spectrochim. Acta*, **21**, 1709 (1965).
26. R. Botter, *Adv. Mass Spectr.*, *Proc. 2nd Conf.*, *Oxford*, R. M. Elliot, ed. Pergamon, Oxford, **2**, 540 (1963).
27. A. Foffani, S. Pignataro, B. Cantone and F. Grasso, *Z. Physik. Chem.*, **45**, 79 (1965).
28. F. I. Vilesov and B. L. Kurbatov, *Dokl. Akad. Nauk. SSSR*, **140**, 1364 (1961).
29. N. A. Beach and H. B. Gray, *J. Am. Chem. Soc.* **90**, 5713 (1968).

Index

(Entry for photoelectron spectrum indicated by S)